Tales of Love and War
from the Mahabharat

Tales of Love and War from the Mahabharat

Retold by

G. D. KHOSLA

Delhi

Oxford University Press

Bombay Calcutta Madras

1994

Oxford University Press, Walton Street, Oxford OX2 6DP

Oxford New York Toronto
Delhi Bombay Calcutta Madras Karachi
Kuala Lumpur Singapore Hong Kong Tokyo
Nairobi Dar es Salaam Cape Town
Melbourne Auckland Madrid

and associates in
Berlin Ibadan

Illustrations by Shouma Banerji Kak

Typeset by Imprinter, C-79 Okhla Phase I, New Delhi 110020
Printed by Rekha Printers Pvt. Ltd., New Delhi 110020
and published by Neil O'Brien, Oxford University Press
YMCA Library Building, Jai Singh Road, New Delhi 110001

Contents

Contents

Acknowledgement

In retelling the stories in this volume I have relied entirely on the text of the ten-volume Hindi translation of our great epic, printed and published by the Indian Press Ltd., Allahabad.

G. D. KHOSLA

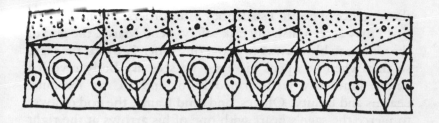

Shakuntala

The sage Vishvamitra (meaning 'friend of the universe') was performing austerities and was endlessly reciting sacred hymns in his resolve to strive for elevation to paradise, to swarga where there is no pain and suffering, no wickedness and untruth, no jealousy and animosity. In swarga all is joy and peace, goodness and love and, of course, eternal life uninterrupted and unhampered by the call of Yama, the god of death.

But the gods who reside in swarga jealously guard their exclusive enjoyment of the things which pander to their physical, spiritual and emotional needs. They get worried if a mere mortal aspires to invade their realm. Thus they began to think that Vishvamitra posed a real danger, for he would intrude into their well-guarded preserve.

Lord Indra was the most concerned of all the gods. To defeat Vishvamitra's designs he summoned Menaka, the most glamorous nymph of his court, and ordered her to go at once to Vishvamitra's hermitage and seduce him by a shrewd display of her charms and by the skilful practice of her allurements. When thus disturbed and swayed from the path of self-denying austerity, the ambitious sage would cease to be a threat to the exclusive glories of the Indra sabha, the court of the gods in swarga.

Menaka feared Vishvamitra's terrible wrath and pleaded that he would destroy her. She appealed to Indra: 'You know the miracles he has performed. His mouth can exhale fire, his blazing eyes are like the sun and the moon and his tongue can pronounce

the sentence of death. I shall obey your orders, but help me so
that I may accomplish my task safely. Send Marut, the god of
the winds, to blow away my covering when I go near the sage,
so that when he opens his eyes, he sees my body revealed
before him—my well-formed limbs and all my feminine
charms and beauty. Oh yes, and send Kama, the god of love,
to pierce the sage's heart with one of his arrows at the right
moment.'

'So be it', said Indra, pleased with himself and with his
favourite apsara. In his mind he saw the complete success of his
design and the failure of Vishvamitra's ambitious adventure.
And, indeed, that is exactly what came to pass.

Vishvamitra just could not resist Menaka's seductive proximity.
Jolted out of his meditation and concentrated study, he took
Menaka for his earthly companion and sported many months
with her in the peaceful seclusion of his hermitage. Menaka
gave birth to a beautiful baby girl, and, her objective accomplished,
she deposited the baby near the banks of the Malini,
after which she went back to Indra's court.

The birds of the forest were attracted by the baby's presence
and hovered round her, chirping and singing soothing songs
till the rishi Kanv, whose hermitage was nearby, discovered
her and carried her home. He brought the baby up as his own
daughter and gave her the name Shakuntala, which means 'protected
by birds'.

Shakuntala grew up in the pious surroundings of her adoptive
father's ashram. She inherited Menaka's beauty, and Vishvamitra's
determination and fiery spirit.

The years passed. Then one day, Dushyant, the young and
handsome ruler of Hastinapur, was out on a hunting expedition.
He was accompanied by his usual retinue of soldiers, chariots,
elephants, and cavalrymen. In the course of his hunt he travelled
many miles and killed many animals. Thus he came to a forest
where there were many hermitages, wherein dwelt sages and
ascetics. In this region the trees were well looked after, there was
an abundance of fruit and sweet-smelling flowers, and bowers
that invited a quiet peaceful sojourn. As he proceeded on his way

he came to the river Malini, and to the hermitage situated near it. Here, the trees and the flower-beds were even more beautiful; birds sat chirping and whistling on branches laden with fruit; peacocks called loudly. Groups of priests sat round holy fires chanting sacred hymns. Dushyant was at once astonished and happy at what he saw and heard. This was Kanv rishi's ashram.

Dushyant left his retinue, took off his royal regalia and assumed a devout and respectful expression. Accompanied only by his minister and his head priest, he entered the hermitage garden to offer his obeisance to the sage. At the door of the dwelling he asked his minister and priest to wait outside, and went in alone, calling out in his loud and authoritative voice, 'who is here?'

Kanv was not at home. Shakuntala, dressed in a simple prayer dress and looking like the goddess Lakshmi herself, exuding heavenly beauty, got up and stood before Dushyant. 'Welcome to rishi Kanv's ashram', she said, joining her hands in respectful salutation. 'Please sit down'. She brought flowers and fruits and cool fresh water from the stream in a vessel. She placed these before him so that he might refresh himself and shed the fatigue of his hunt.

'I wish to see Kanv', said the King, 'Where is he?'

'He has gone to collect fruit', Shakuntala answered, in her simple manner. 'Would you like to wait for him?'

The king was completely bewitched by Shakuntala's beauty and naïve ways. Since he also wished to offer his salutations to the rishi, he readily agreed to stay. He asked her how she came to be living in the hermitage. Shakuntala said she was Vishvamitra's daughter, but had been brought up by Kanv. She narrated the story of her birth. She spoke with charming simplicity, causing Dushyant to be more and more enamoured of her. He resolved to make the beautiful girl his bride and queen.

When she finished speaking and Kanv had still not returned, Dushyant made her a proposal of marriage. He explained that he and she were both Kshatriyas, and that there were six distinct forms of marriage which were legal and sanctioned by custom and sacred precept for the union of a Kshatriya man and woman.

In particular, the 'gandharva' form was at once available to them. 'Do not be afraid', he said, 'and doubt not that a marriage performed according to gandharva rites will be binding on the spouses for life. Tell me what I can do to win your consent. I offer you gold necklaces and ear-rings, the priceless gems of every country, jewels, clothes, and carpets. I shall give you all my wealth, as also my kingdom. O beautiful one, give your consent and be my wife.'

Shakuntala was greatly moved by the king's speech, and, after a moment's pause, she said: 'If this indeed be the path of dharma, and if I have the right to give myself to you, I agree to the gandharva mode of union with you. But I must stipulate that the son begotten of our union will be your heir and will be crowned king after you. This is my firm request. If you grant me this, I shall give myself to you.'

The king, without a moment's thought or hesitation, accepted this condition and promised to conduct Shakuntala to Indra-prastha, the capital city of his kingdom.

And so the union of the young and handsome king and the beautiful maiden was consummated in Kanv's holy hermitage. As he was leaving, Dushyant reassured Shakuntala that he would at once despatch his army, consisting of chariots, elephants, cavalrymen and foot soldiers, to escort her to his kingdom. A thought passed through his mind, 'What will Kanv say when he hears of all this?'

When Kanv returned home a little later, he noticed that Shakuntala had a shy and abstracted look. He guessed that something unusual had happened during his absence 'Don't worry, my child', he said, 'If a woman in love accepts a man also in love with her, she does nothing wrong. Whether she marries him according to vedic rites or enters into a union in the gandharva way, she is blameless and has not transgressed the precepts of dharma. She remains virtuous and honourable.'

Shakuntala took the bundle of wild fruit from the sage and washed his hands and feet. After a little while, when Kanv had rested, she said to him, 'Father, I have Dushyant for my husband. Pray give your blessing to him and to his ministers.'

Kanv thereupon pronounced his blessing, expressing satisfaction with her and the purity of her love for Dushyant.

In due course Shakuntala gave birth to a handsome, bright-eyed baby boy. Kanv performed the rites prescribed by the Vedas and prayed that he would grow to possess the virtues worthy of a king's son. As the boy grew, he was taught the use of arms and all the military arts. He perfected himself in the sport of hunting wild animals, and he acquired the royal virtues of kindness, magnanimity, and respect for both elders and the pious. He was gifted with pearly teeth, lustrous looks and a strong, well-proportioned body. He could ride any horse, climb trees, and, single-handed, overcome wild beasts. He was the delight of the hermitage inmates, who gave him their love and their respect. They conferred on him the sobriquet 'Sarvadaman', which means 'one who subdues all'.

Six years went by in this manner, and yet there was no news of Dushyant. No messenger from him came to seek tidings of the wife he had left behind; no military escort came to fetch her and the young prince. Yet Shakuntala was patient. She kept hoping for the fulfilment of the promise made by her departing husband.

At last Kanv said: 'The time is now ripe for the boy's installation as the king's heir apparent.' Calling a number of his disciples, he told them to escort Shakuntala and the young boy to her husband's palace. 'It is not proper,' he declared, that a married woman stay for so long in her father's home, for thus her happiness, her virtue and righteous conduct may get blemished.'

So Shakuntala and her son were conducted to Hastinapur and led into Dushyant's presence. The king sat before her in all his regal splendour. After introducing her, the disciples returned home to Kanv's ashram.

Shakuntala greeted the king with all due respect and humility, saying: 'O king, here is your son, endowed with all the celestial splendour and princely virtues. Fulfil the promise made by you when you took me as your wife, and install him as your heir apparent.'

Shakuntala looked at the king expectantly in silence. The king returned her gaze, and for what seemed a timeless moment, did not open his lips. What he saw was a beautiful young woman standing before him dressed like a poor ascetic who possessed a strange dignity. Her lips were slightly parted, her shining eyes wide open, and her right hand held the young boy's hand. The innocent smile on the boy's face completed a picture of calm composure, like that before a storm.

The king turned his eyes to the long row of courtiers dressed in gold and silver finery standing in respectful silence. Slowly, he brought his gaze back to the two figures before him. He then spoke in a firm and clear tone empty of all traces of emotion.

'Wicked woman, whose wife are you and whose son is this boy? I have no recollection of ever setting eyes on you or making you any promise. I have had no association with you, neither spiritual, emotional, nor material. So take yourself away or stay as you please, but I do not wish to say anything more.'

Like a pillar of stone, Shakuntala stood petrified. She was overwhelmed by a paralysing sense of shame, mortification, and anger. Then, after some moments, the inordinate strength and the fiery temperament she had inherited from her father, Vishvamitra, together with the conviction of the righteousness she had learnt from Kanv, burst forth like a volcano.

'Your Majesty, you know everything. You remember everything. And yet you behave like a lowly imbecile and pretend not to know anything, not to remember anything. How can you speak like this? In your heart you know what is true and what is false. By speaking so, you are bringing your inner being into contempt. Your soul is witness to what is true. He who utters a lie is capable of committing any sin. He is no better than a thief or a robber. You think that you alone are aware of what really happened. You are mistaken. In your heart dwells the all-knowing Narayan, the supreme god who knows and witnesses all your misdeeds. He who does wrong thinks that no one has seen him do it, but he is wrong. The gods observe everything, and the omniscient Narayan, a part of his own soul, is witness to all the lapses and inequities perpetrated by him. The sun, moon, fire, wind, sky, earth, water, death, day, night, dawn, evening, and dharma, the god of righteousness, are watching and observing every movement of the lives of every human being. The man who pleases the gods and his own soul is given freedom from the cruel bondage of earthly existence, and from the tyranny of the angel of death. But the man who deceives his own soul and professes to be different to what he really is, gets no peace. I am a chaste and virtuous wife. I have come here of my own free will. Do not humiliate me, give me my rightful place. Why do you insult me in the presence of all these courtiers, O King? Remember, a wife is really a part of her husband's body and soul. She is the foremost of his well-wishers. She is the source of his spiritual, material, and emotional being, of his thoughts and deeds. Through this she makes his spiritual salvation possible. She contributes to his happiness and good fortune. A sweet-spoken wife is like a friend who gives joy. She is like a father who guides him on the path of righteousness. She is like a mother who gives him succour when he is sick or afflicted with sorrow. Her company relieves the fears and dangers of a thick and lonely forest. He that has a wife is trusted by everyone. It is for these reasons that the marriage union was ordained. A man, torn by grief or suffering from a painful disease, is soothed and comforted by the companionship of his wife, just as one who is hot and perspiring is refreshed by a cool bath. No one, not even when angry or distraught, should speak harshly to his wife.

'A wife is like a sacred field in which something sown gives birth to its own likeness, and this second self—the son—gives the husband a sense of fulfilment and true joy. What happiness is greater or more precious than that which the father experiences on first seeing his son playing and rolling in the dust and then running to embrace him. This boy, standing by my side, is your son. He is looking at you with wistful expectation, longing to climb onto your knees. Why don't you reach out to him and press him to your bosom? A son's embrace gives ease and ecstacy to the heart. The finest apparel, the touch of the most beautiful woman, the feel of cool water, are as nothing compared to the closeness of a son. Do you know which vedic mantra is recited at the moment of consecrating a son? It is this: O son, your limbs have been created by my limbs, your heart from my heart. Only in name are you separated from me. Truly you are my own body and soul. May you live for a hundred years. You are the saviour of my life and of my race. So may you live for a hundred years . The boy standing before you here is your son, part and parcel of your very being.'

Shakuntala paused for a moment, as if watching for a reaction to what she had said on Dushyant's face. But the king remained silent and impassive. Shakuntala took a deep breath and continued. 'You came to me when you were hunting in the forest near my father's hermitage. When you took hold of my hand, I was a virtuous virgin. Of the six celestial nymphs—Urvashi, Purvachitti, Sahjanya, Menaka, Vishvachi, and Ghritachi— Menaka, the daughter of Brahma is deemed to be of the highest rank. Upon descending from swarga, from paradise, she made love to Vishvamitra, and gave birth to me. But after my birth she cast me away as if I were not her child. What wrong had I done that I was abandoned by my parents and why am I now being disowned by my husband? I shall go back to my father's ashram, but first you must acknowledge your son and keep him here with you.'

Shakuntala drew herself up, exuding from every pore her innate pride and confidence. There was now not the faintest trace of dejection or defeat on her face, which was flushed deep red with righteous indignation, having cast off her initial sense of humiliation. Her demeanour had assumed a determined, almost triumphant aspect as she waited for the King's reply.

'Shakuntala, I have no knowledge of having fathered this lad', he said in an angry tone. 'The utterance of falsehoods is a womanly characteristic. Who can believe her words? Menaka, devoid of the capacity to love, begot you. She discarded you as one discards the flowers offered to a god once the prayers are over. And who is your sire? The lustful Vishvamitra, a Kshatriya turned Brahmin who is quite innocent of lasting love for anyone. But even if Menaka is the greatest celestial nymph and Vishvamitra the best of sages, why do you, being their daughter speak like a wanton woman and try to ensnare me?

Are you not ashamed to address me in this manner? Take your-self away. You are the offspring of a lustful seducer and have come to me in the garb of a holy ascetic. I do not know you. Go where you please, but leave me.'

Shakuntala answered with dignity. 'Your Majesty, you are ready to see the faults of another, be they as small as the grains of a mustard seed, but you overlook your own faults, though they be the size of a bel fruit. I was mothered by Menaka, the foremost of heavenly nymphs, and my blood is of a higher order than yours. You are able to walk on earth only, whereas I can roam the upper spheres of the universe. An ugly man believes himself to more handsome than all other men until he faces a mirror. But he who is truly handsome does not ridicule others. As swine always wallow in slime and dirt, preferring this to a bed of sweet smelling flowers, so the wicked take pleasure in talking ill of others. They show little regard for the old and the infirm. They disdain to tread the path of virtue. The good, on the other hand, see only virtue in others and overlook their misdeeds. A Father who disowns his son will be destroyed by the gods and will never attain salvation. The son carries on the family line and his father's race. Manu has said that a son embraces both the virtues and the worldly possessions of his father and thereby adds to his happiness. It therefore ill behoves you to denounce your son. It has been said that there is no sin greater than uttering falsehood, so do not destroy your virtue by killing the truth. Sir, if you do not believe me, I shall go back to my home in the forest. Indeed, I have no desire to stay with a man like you. But take warning, this my son will one day rule the entire earth.'

Shakuntala turned and began to walk away from the king who had remained all this while seated on his throne. Dushyant now rose, saying in a loud voice, as if inspired or possessed, 'Stop, O beautiful one, stop, my beloved, my lawful wife and the Queen of my realm. Look, I remove the mask of pretence. The gods in heaven wish it. Seeing your distress they have commanded me to end your trial and your tribulation. So, I now openly acknowledge you as my wife and the child begotten by you as my son, for he is a portion of my own self, both in body and in spirit. I pretended to be ignorant of who you were and that you and my son were strangers because I wished to convince my courtiers of the truth of the matter. After hearing you, they are left with no doubt that you have spoken the truth and that you are indeed my lawful wife and this lovely child is indeed my son. I atone for my wrongs by openly greeting you, most beautiful and virtuous woman, and acknowledge you as my wife and my Queen, and your son as the rightful heir to my kingdom and all its possessions.'

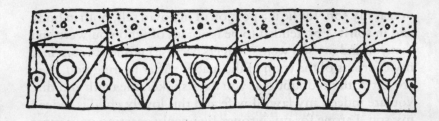

Ghatotkach

Fleeing from the burning house of lacquer, the five Pandav brothers and their mother, Kunti, hastened through forests, across streams, and over mountains in order to throw Duryodhan off their scent.

By nightfall they arrived in the middle of a thick forest far from any human habitation. They were very tired and thirsty, and Kunti repeatedly asked for water to slake her thirst. Bhim was extremely distressed by his mother's condition, and said he would go and look for water while the others rested under a tree. He walked briskly, looking for a stream or pond, and soon came to a clear-water lake, where he quickly refreshed himself. For lack of any utensil in which to carry water for his mother and brothers, he soaked his clothes so that they were full of water and hastened back to where he had left then. When he returned he found them all fast asleep, so he decided to sit and keep vigil till they awoke.

Nearby, in a thick sal tree, there lived a predatory man-eating monster, Hidimb, and his sister, Hidimba. The scent of humans began to tickle Hidimb's nostrils, and he cast his big red eyes around to trace the source of the odour. Soon he saw the five humans lying down under the tree. This sight so pleased him that he began scratching his head with his long pointed nails and his mouth watered in anticipation of the tasty meal he was about to consume. He relished human flesh more than anything else. So he asked his sister, Hidimba, to go and kill these five people and to bring them to him so that he could enjoy his meal. 'Go quickly', he urged, 'and let me satisfy my urge to taste human meat.'

Hidimba walked quietly to the place where the Pandavs slept. The moment Hidimba saw the broad-shouldered Bhim with his unrivalled masculine handsomeness, she fell madly in love with him and thought, 'This wonderful man is fit to be my husband. I shall never carry out the directions of my wicked brother: a woman loves her husband and not her brother. What shall I gain by killing him, only a moment's pleasure as I eat his flesh., But if he lives I can spend an eternity living with him, loving him.'

So, Hidimba promptly conjured up supernatural powers and assumed the form of an extremely beautiful young woman. She walked with slow, alluring steps toward Bhim. When Bhim raised his eyes, he saw a glamorous young girl with tantalizing, full, rounded breasts and large doe-like, gentle eyes. Smiling shyly at Bhim, Hidimba said to him, 'Respected sire, who are you? Where have you come from? Who are these four god-like men and who is this beautiful woman with the golden complexion and soft, shapely limbs? They are all sleeping so peacefully. Don't you know that demons live in these woods? This is where the wicked Hidimb lives. He is my brother and he has sent me to you to kill you and to take your bodies to him for his meal. But believe me, after seeing you, all I want is to have you for my husband. I am truly and wholeheartedly in love with you. Make me your wife, I beg you. I shall save you from my demon brother, and we can run far away from here and spend our lives loving each other and sporting together.'

But Bhim did not wish to abandon his brothers and his mother and run away with Hidimba, so he declared to Hidimba that he would stay and fight Hidimb, saying that he had enough power in his limbs to overcome any demon.

Hidimb had, by now, descended from his sal tree, and impatient at his sister's long absence, walked toward the place where the Pandavs were resting. When he saw Hidimba transformed into a most beautiful woman he became exasperated, and swore that he would kill and devour all six humans and then kill her as well, for she had disgraced the entire tribe of demons by wanting to marry a human.

His eyes inflamed with rage, and grinding his teeth, Hidimb advanced to attack the Pandavs. Bhim shouted, 'Stop, don't disturb the peaceful sleep of my brothers—and why are you thinking of injuring this beautiful woman? It is not her fault that she proposed marriage to me. Kama, the god of love, moved her heart and compelled her to desire me. And remember that you yourself had asked her to come here. As long as I am alive I will not let you lift your hand against her.' Saying this, he quickly took hold of Hidimb's hand, dragged him to some distance away, and began to hit out at him with his hands and feet. The demon fought back with all his might. Thus, a long, gruelling and bloody battle started.

The two combatants were almost equally matched. Hidimb called forth all his demon strength, trickery and magic, while Bhim, who was strong enough to uproot whole trees and hurl them at his adversary, sought to establish his superiority. Seeing his chance at one stage, Bhim lifted Hidimb bodily above his head, and spun him round and round, hurled him to the ground and then strangled him. Bhim then bent Hidimb backwards so that his back broke. Upon hearing the noise of the fight, the Pandavs had woken up to see the beautiful Hidimba standing next to them and Bhim fighting Hidimb some distance off. Seeing the demon completely overcome and slain, they rushed to embrace their brother and congratulate him and then heard the whole story from him.

The fear of discovery by Duryodhan's agents still haunted them, however, and so they decided to quickly leave the place. When Hidimba started to follow them, Bhim advised her to stay behind lest the demons, on coming to know of Hidimb's death and finding her gone, should persue and harm her. But Hidimba persisted in going with Bhim and told Kunti, Bhim's mother, that she must accept her as Bhim's wife and her daughter-in-law. Kunti saw Hidimba's genuine devotion to Bhim and so gave her consent to the union.

And so, Bhim and Hidimba were married and, in due course, Hidimba gave birth to a strong and valiant son who was named Ghatotkach because his head resembled a pitcher and the hair of his head was thick and stood up on end.

Eventually, over the course of time, the Pandavs decided to leave Hidimba and her son. So, it was agreed that whenever the Pandavs needed Ghatotkach's assistance, they had only to pronounce his name and he would appear at once to do their bidding.

Indeed, it was not long before the Pandavs did need Ghatotkach's help. When climbing a steep mountain, they became exhausted by the long and arduous effort. As soon as they remembered Ghatotkach and wished that he were present to help them, he instantly appeared before them and carried Kunti on his shoulders up the mountain.

Ghatotkach's valour and his love for the Pandavs was made truly manifest, however, in the battle of Kurukshetra, in which he fought day after day and performed feats of great bravery and endurance. On the very first day of the battle, he fought with Alambush, a greatly dreaded demon-king who was an ally of the Kauravs. Ghatotkach hurled no less than ninety sharp, pointed spears at his adversary, wounding him in many places. He, too, received many bloody injuries at Alambush's hands, but the encounter remained inconclusive, and each retired to rest and recuperate.

On the fourth day of the great battle, Bhim was causing such panic among the Kaurav forces with his ceaseless rain of arrows and spears that, hoping to stem the tide of Bhim's onslaught, fourteen sons of Dhritarashtra made a concerted attack upon him. Bhim very quickly slayed eight of them. The remaining six ran away, uttering cries of despair. To avert further disaster, Bhishm urged his warriors to also go and attack Bhim. So, Bhagdatt rushed towards Bhim, shooting arrows at him in rapid succession. However, Bhim's supporters surrounded Bhagdatt and attacked his elephant with spears, but the elephant only quickened his pace as it neared Bhim and Bhagdatt pierced Bhim's chest with his spear. Bhim fell, and swooned, clutching at his chariot. Seeing him fall, Bhagdatt shrieked with joy.

Suddenly, Ghatotkach rushed up as if from nowhere and attacked Bhagdatt. Ghatotkach was astride a mighty elephant, resembling Aravat of Lord Indra. Urging it forward to kill Bhagdatt, he managed to wound him. The Pandav forces were heartened by the presence of the mighty Ghatotkach and seeing his indomitable courage, advanced the attack. Bhishm saw that it was therefore advisable to cut his losses and ordered his men to retreat. Thus was Bhagdatt saved from Ghatotkach's fury. Bhim, in the meantime, was attended to. He later quickly recovered from the injury. There was great rejoicing in the Pandav camp over their success in the battle of the fourth day.

Ghatotkach and Bhagdatt met again on the seventh day of the battle. This time, Bhagdatt was riding a mighty elephant. Ghatotkach advanced upon him in a chariot. Seeing Ghatotkach going towards the enemy in a fearless, determined manner, the Pandav warriors took heart and rallied to support their demon leader.

Ghatotkach showered his arrows on Bhagdatt like a torrent of rain. He hurled no less than seventy spears at Bhagdatt and inflicted heavy wounds on his elephant. But Bhagdatt and his mount remained undaunted and Bhagdatt succeeded in killing all four of Ghatotkach's chariot steeds. At this the demon leader lost heart and retreated from the battle, resolving to resume his offensive on the following day.

As the sun was setting, Bhagdatt rode forward, his elephant trampling underfoot scores of Pandav worriors.

The next day's encounter again ended indecisively. Bhim slaughtered sixteen more sons of Dhritarashtra, but he lost his own son, Iravan. This provoked Ghatotkach to spring into action once again, attacking Duryodhan. Bhagdatt then came to Duryodhan's defence and the fierce combat went on for a long time. Many warriors on both sides met their deaths, the mutilated bodies presenting a gruesome spectacle as the day ended. Duryodhan, Bhagdatt, Bhim and Ghatotkach, however, returned to their respective camps.

Day after day, the hostilities went on relentlessly, taking a heavy toll of the brave, valiant men on both sides, though the leaders of the Kaurav and the Pandav forces continued to put heart into their men. Thus the war entered its fourteenth day, the battle raging fiercely throughout the day. When the sun set, no one thought of resting or returning to camp. Torches were lit by the men of both sides and the combat went on.

Karan played havoc wherever he went, and the Pandav warriors were beginning to lose heart in the face of his indomitable offensive. Seeing this, Yudhisthir spoke to Arjun and urged him to take steps to slay Karan. Arjun at once declared that he would himself engage Karan in personal combat. Arjun therefore asked his charioter, Krishna, to drive him close, but Krishna counselled him against this step and said, 'There are only two persons who are capable of standing up to Karan, you and Ghatotkach, but at this juncture I do not think it advisable for you to fight with him. He has with him the dreadful weapon of Shakti which lord Indra himself gave him. He has kept it specially to destroy you, so at this stage, your well-wisher Ghatotkach should go and oppose Karan. He was begotten by Bhim and has more than his strength of body and can conjure up supernatural weapons. Look how deftly he disposed of Alambush's father only this morning, despite his ability to invoke demonic weapons. So, send him to deal with Karan.'

Arjun sent for Ghatotkach and when the dark and fearsome son of Bhim arrived Krishna said to him, 'Ghatotkach, listen carefully to what I say, the moment to display your true valour has arrived. Save you, no one can stand up to Karan. You have command over a host of demonic weapons. The Pandav army is about to be swamped by the flood of Karan's might. Become their lifeboat, and save them from drowning in the torrent of despair. Be the truly righteous son to Bhim. It is to cope with such situations that men desire to beget sons and it is the duty of a son to defend his father and his elders. No one but you can stand up to Karan and stop the carnage he is inflicting on your army. Demons are at their best during the hours of darkness, and at this time, everything is in your favour. So, go and engage Karan in single combat and annihilate him.'

Ghatotkach was delighted to be given the opportunity to

display his valour and declared that there was no one in the whole of the Kaurav army whom he could not defeat.

Ghatotkach then rushed to where Karan was standing and at once engaged him in a fierce battle. The story of his duel with Karan will be related in glowing terms till the end of time.

Alambush, the demon whose father had been slain by the Pandavs, ran to avenge his father's death and attacked Ghatotkach. A fierce battle ensued but Ghatotkach prevailed over his assailant and succeded in mortally wounding him. He then threw him on the ground and severed his head from his body. He carried the head by its hair and in a gesture of triumph, threw it in front of Duryodhan as he sat in his chariot. Then Ghatotkach roared at Duryodhan, 'Look how I have slaughtered your ally, I shall now go and deal with Karan in the same manner. It is written in our scriptures that when one goes to visit a king, a Brahmin or a woman, one should not go empty-handed. That is why I have brought you Alambush's head as my offering to you. You can sit here till I slay Karan.'

Ghatotkach then rushed towards the spot where Karan was standing, sending a veritable shower of arrows at him. Ghatotkach presented a fearsome aspect wielding his powerful bow: his eyes inflamed with wrath, his massive body, long and powerful legs, his huge head towering above his broad shoulders. His appearance struck terror in Karan, who turned to meet his adversary, shooting arrow for arrow. It was like one elephant encountering another, or two bulls charging at each other. Each drew their bow-strings right to their ear, releasing their mighty arrows with full force. The arrows of each pierced the armour of the other, wounding both of them in countless places, streams of blood flowing from both combatants.

The fight continued relentlessly on into the night. Alayudh arrived, a demon who nursed an old grudge against the Pandavs and began fighting Ghatotkach. It was a fierce and bloody struggle in which Ghatotkach was victorious. In victory, he threw Alayudh to the ground and cut off his jewel-adorned head.

With Alayudh's death, the duel betweeen Ghatotkach and Karan was once again resumed. Ghatotkach was now even more

aggressive than before. He slaughtered the Kaurav warriors
one by one, all the while keeping Karan at bay. He then killed
all four steeds of Karan's chariot, forcing Karan to leave the
chariot and to worry about his next step. He saw that the Kauravs
were losing heart and stood petrified by the murderous onslaught of the fearsome demon. Some of the Kaurav warriors
shouted to Karran to use the Shakti weapon which Indra had
given Karan in return for his gift of a pair of divine ear-rings.
'Now is the time', they pleaded, 'to use this weapon and save
yourself and the entire Kaurav army from the danger of total
annihilation.'

Karan had for many years kept the divine weapon for use
against Arjun, whom he had resolved to kill. But the turn taken
by these events made it imperative that Ghatotkach should be
eliminated immediately, otherwise the battle of Kurukshetra
could be irretrievably lost. However, the weapon could be used
once and once only, for on use it would destroy the person at
whom it was aimed and then fly back to Indra in his celestial
abode. But Karan saw no alternative, and taking hold of Shakti,
hurled it at Ghatotkach.

Thus ended the glorious tale of Ghatotkach's valour, devotion
to duty and intense filial love. In death, as his gigantic body lay
on the battlefield, crushing under it a whole army of Kauravs,
he looked even more formidable than before. His heroic end
was marked by the blare of conches, the beat of drums, the
clanging of cymbals and the shouts of the Pandav warriors.

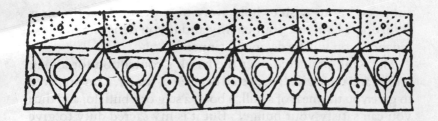

Hospitality

King Shivi was a renowned ruler from long ago. His large-hearted hospitality and whole-hearted solicitude for the safety and comfort of those who sought asylum in his domain were known to all.

On one occasion, as he sat in the palace garden, attended by his ministers and courtiers, a pigeon who was being pursued by a hawk, suddenly flew down and dropped into his lap. King Shivi spoke to him in a comforting tone, 'Don't be afraid. Tell me how and by what you have been frightened. You have come to me, so now you need have no fear. Look, here are all these people to protect you against all kinds of danger. No one will dare harm you. For your protection, I am ready to spend all my wealth, and if need be, my very life will be readily given.'

As he was speaking to reassure the pigeon, the hawk arrived and said to the King, 'Sire, this pigeon belongs to me. I got him with great effort. Do not meddle in this affair. I wish to satisfy my hunger by eating the pigeon. Do not deprive me of my food. I have already left my mark on him with my beak and talons. You can see the blood oozing from the injuries I inflicted on him. Mark how he breathes with difficulty. I am famished, and I must have him for my dinner. Don't try to shield him or protect him. You are the lord of human beings. Your sovereignty does not extend to the hungry and thirsty birds of the sky. You can exercise your jurisdiction over your friends and foes, over the members of your family and your subjects, but to wish to hold sway over the birds of the air is improper. I am not your enemy and if you deny me my sustenance, you will be deemed a sinner.'

The King pondered over the hawk's contention for a long while, and then said, 'Respected denizen of the air, I am prepared to give you the meat of a bull, a boar, a stag or a buffalo, and thus you can satisfy your hunger. But it is my sacred duty to give protection to one who seeks asylum in my domain. I cannot hand the pigeon over to you to be devoured.'

The hawk rejected the King's offer, saying, 'I do not eat the flesh of a bull, boar or stag. God has ordained that the birds of the air will be my food. Everyone knows that hawks live entirely on pigeons, sparrows and other birds. If you have so much love for the pigeon, you can give me an equal measure of your own flesh.'

King Shivi promptly replied, 'You have done me a great favour by expressing your true desire. I shall at once give a you a quantity of my flesh, equal in weight to the pigeon.'

Shivi then sent for a pair of weighing scales, and placing the pigeon on one side, began cutting pieces from his limbs, putting them on to the other side of the scales. Hearing of this strange happening, the womenfolk of the King's subjects came, wailing and lamenting. The cries of his ministers and courtiers filled the palace halls. The sky became clouded over and the earth trembled. The King continued to cut off flesh from his hands and feet and legs and arms, till gradually the flesh of his entire body had been heaped on to the weighing scale. But the scale with the pigeon's body on it still remained the heavier one. Finally, when only his bare skeleton remained, he climbed on to the scale himself.

As soon as he did this, Lord Indra descended from his celestial abode and manifested himself. He and the other deities attending upon him showered flowers on Shivi and sounded trumpets of glory. Celestial nymphs performed a dance of joy to celebrate his righteousness, his act of true virtue, and his undaunted hospitality.

Shivi was then immediately restored to his original self and continued to rule over his people for many years.

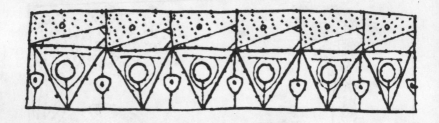

Savitri and Satyavan

This is a story from many years ago when men trod the path of virtue and women were chaste and dutiful. In those days, the people of Madra were ruled by King Ashvapati whom none could surpass in righteousness or the just administration of his realm. He paid due honour to the Brahmin priests who came to his court. He led a pious life and acted with consideration and generosity in dealing with his subjects. His queen, Malvi, was the embodiment of purity and chastity. She performed her royal duties with unswerving devotion, her mind ever intent on the well-being of her royal consort and the precepts of dharma. But amid all the affluence and pomp and show of their regal establishment, they were living with a deep sense of deprivation because of the lack of the greatest blessing that the gods confer on men. They had no son who could gladden their hearts in their youth, or be their companion and support in old age, or who would, after their death, succeed to the royal state and minister to their souls by performing the rituals and prayers prescribed in the Vedic texts.

Ashvapati thus reached the age when men are too old to beget offspring and he felt the burden of sorrow and mortification oppressing him beyond endurance. To relieve his suffering, he resorted to a long and arduous course of austerity and prayer. He subjected himself to the most rigid control of his desires. He ate sparingly, repeated the sacred mantras and made repeated offerings to the goddess Savitri, the heavenly consort of Lord

Brahma, the supreme spirit who created the universe and who
is all powerful, all pervading. Ashvapati continued to pay unre-
lenting homage to Savitri for eighteen years. At the end of the
period, Savitri emerged from the sacrificial fire and made herself
manifest. He saw her standing before him in all her celestial
glory. She said, 'O King, I am well pleased with your devout
prayers and the austerities you have imposed on yourself. Ask
for any gift from me and it shall be granted. But do not ever
deviate from the path of virtue.'

'I pray for the blessing of sons. Thus I will be able to
continue treading the path of virtue. The holy Brahmins have
advised me that only by begetting sons can I safeguard the
continuation of my race and my own spiritual salvation.'

Goddess Savitri replied, 'By my power of prescience, I knew
that that was your heart's deep desire. I have spoken to Lord
Brahma regarding the matter. He is pleased to grant you the
blessing of a bright and glorious daughter. You must be content
with this and say no more about your longing to have a son.'

In course of time a lotus-eyed baby girl was born to queen
Malvi. She had all the beauty of the goddess Lakshmi. After
her birth, the King recited the Gyatri mantra which is also
known as Savitri mantra and gave her the name Savitri. The
child grew up, burgeoning into a maiden resembling a celestial
nymph, with slender waist and rounded hips. Such was her
physical perfection and the splendour of her beauty that no one
had the courage to seek her hand in marriage.

One day, after taking her bath and saying her prayers, Savitri
put on her finest clothes and, taking a garland of flowers, went
to her father. After offering him the garland, she stood before
him with folded hands. Ashvapati, seeing her standing before
him like a goddess, said, 'My child, you are old enough to
marry, but no one has so far come to ask for your hand. I think
you should yourself select a suitable bridegroom. When you
see someone who pleases you, come and tell me. I shall, after
due consideration, give you to him in holy matrimony. Let me
tell you what the learned priests have told me. According to
our sacred texts, the father who does not arrange the marriage

of his daughter, the man who does not lie with his wife at the appropriate time, the man who does not make proper provision for the welfare of his widowed mother are all to be abhored. So, it is my desire that you find a worthy groom, so that I do not stand condemned and vilified before the gods.'

The King at once ordered that a royal chariot, furnished with trappings of gold and cushioned with silk, be made ready to convey his daughter to countries both near and far. For her escort there was to be a party of old and faithful retainers. So began Savitri's strange adventure.

A great deal of time passed as Ashvapati and his courtiers waited for the return of the princess and the accomplishment of the King's design. Then, one day, as the King sat in his court entertaining the sage Narad who had come to pay him a visit, he saw his daughter, Savitri, walk in. She first went up to Narad and touched his feet in respectful obeisance. Narad thereupon asked the King, 'Where has your daughter been? From where is she coming? Sire, she is now of marriageable age. You should find a husband worthy of her.'

Vishvapati replied, 'That was precisely the errand upon which I had sent her. Hear from her own lips whom she has chosen as her bridgegroom.'

Savitri then related her story. 'Father, there was a king Dyumatsen who ruled over the state of Salva. He was a Kshatriya of great virtue and was blessed with a young son who would have succeeded him. But a god's wrath destroyed his sight. Thereupon, an old enemy invaded Salva and took forcible possession of the blind King's kingdom. The unhappy King, accompanied by his wife and child, escaped from the palace and sought refuge in a forest. Since then, he has been living the life of a poor ascetic. The young prince, named Satyavan, is now a robust, fully-grown man. I deem him worthy of being my husband. In my heart I have already accepted him as my true life companion.

The sage Narad became thoughtful. After a moment's silence he said. 'Savitri does not know the entire truth. She has been too hasty in choosing Satyavan for her husband.'

Ashvapati asked anxiously, 'Is the young prince not possessed of a bright intellect? Is he not valiant, just and devoted to his parents?'

Narad said, 'Yes, yes, he is all that. He is bright like the sun, wise like Brihaspati, brave like Indra, forgiving and tolerant like Mother Earth. In short, he is handsome to behold, magnanimous in spirit, truthful in speech; he is at once valiant, courageous and virtuous; he is not arrogant nor overbearing, but...' Narad paused, knitting his brows and concentrating his gaze on Savitri who was now standing rooted to the ground like a pillar of stone. An unknown fear and foreboding had taken possession of her and rendered her insensible.

Narad turned his head towards the king and, in a quiet, almost soothing tone, went on; 'Yes, the young prince is endowed with all the noble qualities that the gods can bestow, but he has one imperfection which outweighs all his virtues, and this imperfection is beyond the power of anyone to remedy. Yes, one shortcoming only. A year from now, he must die. It is ordained that then his spirit will leave his body.'

Ashvapati sustained the shock with royal fortitude and calm. He at once addressed his daughter, 'Savitri, my beloved child, seek another spouse. Narad has revealed a monstrous imperfection which by far outweighs all his virtues and renders them of no avail. He will be alive for twelve months only and will then leave this world.'

Savitri replied with quiet conviction, saying that a human being dies but once and that a daughter may be given away in marriage by her father but once. 'I have made my choice, now and for all time. I am firm in my resolve to marry Satyavan, whatever be the portion of life granted to him by the gods. I cannot let the very thought of another man being my life companion enter my mind.'

That settled the matter, and preparations for the wedding began without delay. The auspicious day for the ceremony was named by the holy priest. The King called his courtiers and Brahmins to accompany him and his daughter to the hermitage in the forest where the exiled King Dyumatsen lived. Ashvapati greeted Dyumatsen, who was a blind old man seated on a cushion of grass in the shade of a sal tree, and then stated the purpose of his visit. Dyumatsen replied, 'Here we live the harsh life of ascetics. We have been deprived of our royal palace and the comforts it provided. How can your daughter, brought up in the abundance and luxury you have always provided, feel happy in these surroundings, in this rough and ungracious asylum.'

'I and my daughter,' said Ashvapati, 'have made our choice with the full realization of what you say. We know that ease and unease, happiness and sorrow are the lot of everyone. But these things pass. Accept and welcome my daughter as your son's bride, and do not hesitate to receive her into your family.'

Thereupon, the wedding ritual of Satyavan and Savitri was duly performed according to the sacred texts of the Vedas, and they were pronounced man and wife. Ashvapati returned to his palace and Savitri, discarding her costly apparel, donned the ochre coloured clothes appropriate to her new way of life.

In the days and months that followed, Savitri led a quiet and virtuous life, performing her household duties, and tenderly, earnestly, selflessly, providing comfort and happiness to her husband and his aged parents. But all the while, the words spoken by Narad kept reminding her of Satyavan's approaching doom and this gnawed at her heart, day and night.

Only four days remained of the twelve-month period pronounced by Narad. Savitri had kept count of each passing day, and knew that Narad's prophesy was to be shortly fulfilled and that her husband's soul must take leave of his earthly body. So she undertook a total fast for three days. When Dyumatsen heard of the rigorous austerity, he spoke to her anxiously and expressed concern at the suffering that a long fast of three days and three nights must inflict on her tender person. But Savitri was adamant and declared her determination to carry out her resolve. 'Be not apprehensive,' she pleaded, 'only by courage and perseverance can one's objective be achieved.'

Thus the three days passed as Savitri languished and lost weight. Her face assumed the cold expressionless fixity of a wooden doll. She spent the last night in unremitting anguish, and when the sun had risen she began repeating to herself, 'To-day is the day when my lord will die.' She offered her oblations to the sacred fire, and after paying her respects to the Brahmins and her father-in-law, stood before them with folded hands. In response, they uttered a loud prayer beseeching the gods not to let Savitri suffer the misery of widowhood. Her mother-in-law pressed her to her bosom and begged her to end her long and arduous fast and take some nourishment. But Savitri replied, 'My vow will be completed when the sun goes down. Then I shall eat.'

Thus, she remained in anguish, waiting for the hour and the moment when Narad's words would come true. Meanwhile, Satyavan picked up his axe, swung it on his shoulder and pre-pared to go to the forest to gather wood. Savitri cried out, 'You cannot go alone. I shall come with you. I cannot bear to be separated from you today.'

Satyavan spoke to her in soft tones, 'My beloved, you have never been to the forest. The path I have to tread is long and hazardous. The long fast of three days has made you very weak. You will not be able to walk there.'

But Savitri insisted and replied, 'I have not suffered in the least by my fast. The walk through the forest will not tire me, and I have set my heart on accompanying you.'

Satyavan saw the pleading look on her beautiful face, and re-signing himself to her desire, nodded his head in assent. He then asked her to receive his parents' blessings.

As they proceeded through the forest, Savitri's heart was beating fast. All round them sweet sounding birds sang their unending chorus of sylvan music, streams of limpid water and trees laden with multicoloured flowers demanded their atten-tion, but the memory of Narad's fatal words kept nagging at her heart.

At last they arrived at the spot where Satyavan had chosen to hew wood. Here he selected a tree and began to wield his axe, cutting down branch after branch. But he had not quite finished when he began to perspire, and his head began to ache. He said to his wife standing near him that he felt unwell.

'All my limbs,' he complained, 'have lost their strength. My heart is throbbing, my head is being pierced by a dart. I cannot stand. I must lie down.

Savitri sat down with him and placed Satyavan's head on her lap. She began to count the number of months, days and hours that had elapsed since Narad had pronounced her doom.

Suddenly she saw, standing close to them, a monstrous looking being, wearing red clothes, with a shining diadem on it's forehead and holding a rope with a noose at one end. The being was of dark aspect, with blood-red eyes, and he was staring at Satyavan in a fearsome manner. Savitri's heart began throbbing violently. She gently moved Satyavan's head to the ground and slowly stood up. She approached the stranger with folded hands and in a tearful voice said, 'Seeing your formidable presence, I can guess that you are some god and not a human being. Please, tell me who you are and what you wish.'

'Savitri,' answered the dark being, 'You are a devoted wife and a pious woman. So, I can speak to you plainly. I am Yamraj, the lord of death. Your husband Satyavan's time in this world is over and I have come to tie his spirit in this noose and take it away. I would ordinarily have sent one of my assistants to do this job, but Satyavan is no ordinary individual. He is endowed with singular beauty of person and countless virtues. In his accomplishments he has no equal. For this reason, I have come in person to fetch him.'

Saying this, Yamraj dragged out Satyavan's soul, and securing it tightly in his noose, started to walk away towards the South, for that is where the nether world is situated. Savitri followed him. Seeing her coming behind, Yamraj said, 'Go back and perform the funeral rites of your dead husband. You have fulfilled all your wifely obligations. Thus far you follow me, but no farther.'

Savitri answered, 'Sire, wherever my husband goes or is taken by another, I must go. This is what the Vedas have prescribed. Drawn by love for my husband and your kind forbearance I can go anywhere and everywhere. It has been said that by taking only seven steps with another, marital love binds the two together for ever. Let me tell you something about such love. One who is not chaste and virtuous will in vain strive to achieve merit by pursuing the path of asceticism. True knowledge is gained by the diligent performance of dharma, the truly righteous conduct which is a manifestation of wifely love and devotion. It is this true knowledge which the wise have declared to be the ultimate human and godly virtue. And those who acquire such merit need not strive on the four-fold tasks of celibacy, family life, withdrawal from earthly possessions, and total renunciation. Satyavan and I did not resort to a celibate life or total renunciation. Indeed, family life lived with honesty and unswerving righteousness will lead to the ultimate knowledge and wisdom which is deserving of immortality and entry to the celestial sphere. Do not, therefore, impede me in the pursuit of my sacred purpose.'

The lord of death replied, 'Pray, do not persist. I am greatly impressed by the justness of what you say. Your ideas are admirable. The determination and courage you display are unrivalled. I am willing to bestow upon you any gift, bar the life of your husband, as a reward for your virtue and determination.'

Savitri promptly asked for the restoration of her father-in-law's eyesight. Yamraj readily granted the boon and asked her to return home. 'You must be very tired', he said, 'after this long walk. Do not mortify your spirit further.'

But Savitri continued to follow Yamraj, saying, 'I feel no mortification of mind or of body in remaining with my husband, for he means to me the fulfilment of my life's purpose. Sire, I shall go wherever you take my lord. Let me tell you that even a brief encounter with a pious person is a gain, and closer association with him is to be greatly valued. The friendship of a true dharmatma, one who is the soul of virtue, can never be without just reward.'

Yamraj said to her, 'Your words make me happy. They are full of wisdom and piety. Ask of me a second gift and I shall readily grant it, but do not ask for your husband's life.'

Savitri, without a moment's thought, begged for the restoration of her father-in-law's kingdom from which he had been driven out by his enemies. Yamraj readily granted this request also, and admonished her to leave him, for to go farther would inflict even greater hardship on her. But Savitri persisted, protesting, 'My lord, you alone conduct everyone along the path of his or her fate and confer upon each the reward for the deeds performed by them. You alone punish those who are guilty of doing wrong. That is why you are named Yama. Let me add that the truly virtuous do not inflict injury on any one in thought, word or deed. They are generous even to their enemies when they come to seek favours. I am but a weak woman.'

'Just as cool water is welcome to a thirsty man', said Yamraj, 'so is your speech pleasing to me. Ask for yet another gift, anything except the life of your husband, and I shall willingly grant it.'

For her third request, Savitri asked that a hundred virtuous sons be born to her father who had not been blessed with male offspring. Yamraj promptly granted this also, and again asked her to return home. 'You have travelled far', he said, 'and I strongly advise to you to venture no farther.'

But Savitri was not to be diverted from her purpose. She remonstrated, 'My lord, I am with my husband, so the journey does not seem long or harsh. Indeed, my mind is travelling even beyond my earthly body. Listen to me as you go on your way. You are the valiant son of Vivasvat, the bright one. That is why you are called Vivasveta by the wise who know you to be the son of Lord Surya, the sun, who is brighter than all else in the firmament. And since you administer even-handed justice to all, friend and foe alike, you are also known as Dharamraj. A man seeks the support of his friends. It is friendship that inspires trust, and for this reason, a man trusts his friends and I repose trust in you and seek your support.'

Yamraj replied, 'Beautiful one, I have not heard such wisdom-filled words uttered by any other. Your speech is most gratifying. Ask of me yet another gift and I shall happily bestow it upon you, but name not the life of your husband.'

At this, Savitri said, 'Grant me the gift of a hundred sons begotten by Satyavan. This, my fourth request, I beg of you.'

Without thinking, Yamraj, greatly pleased with Savitri's wisdom and humility, promptly granted this also. 'Princess', he said, 'you will give birth to a hundred sons who will be brave and will make you happy. And now go back, for you have really walked a very long way and to go farther will be beyond your powers of endurance.'

But Savitri persisted in following Yamraj, and as she walked, she spoke to him: 'My Lord, the truly righteous are dedicated to virtue. Communion with such people yields good results. It is the truthfulness of the virtuous that makes the sun move in the sky. It is they who support the earth and keep it firmly in position. It is from them that the past and the future proceed. The righteous give without expecting anything in return. So the righteous are the support and the protection of everyone.'

Yamraj turned to her for the fifth time and said, 'You are a dutiful wife full of wisdom and meaningful words. The more you speak the more my respect for you increases. You may now ask for any gift you wish.'

With her heart beating fast and her hopes rising high, Savitri said; 'The gift of a hundred sons which you have already granted cannot be realized without my husband coming alive and being with me. I do not cherish any joy withouut the company of my husband. I crave for neither wealth nor glory without his company. I do not have the desire even to live without him. Give me back my husband. That is my last prayer and supplication.'

Yamraj now granted her request and released Satyavan's soul from the noose. It re-entered Satyavan's body at the spot where it had lain all this time. When Savitri returned to the place where Satyavan had been cutting wood, she saw him move and wake up, as if from a bad dream.

Thus it came about that the deep love of a chaste and dutiful wife gave her father male heirs, restored the sight and the lost kingdom of her father-in-law, and ensured the continuation of her husband's life and his line of successors.

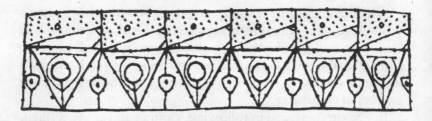

Vishvamitra

The story of Vishvamitra is also the story of Vashisht. Both these illustrious sages lived in the same era, both attached themselves to the court of King Sudas and advised him on the conduct of his government. Both aspired passionately to become the King's chief prohit, a position of considerable importance and power. This circumstance and their rivalry in the achievement of Vedic learning and spiritual wisdom made each intensely jealous of the other. This jealousy manifested itself in mutual hatred. Each was chosen by Brahma to be his mouthpiece in uttering the hymns and verses of the Rig Veda. Vishvamitra thus spoke the hymns of the third mandala which contains the famous Gayatri Mantra, while Vashisht is reported to have recited the hymns of the seventh mandala. Both rishis were hallowed as members of the seven most eminent sages—the group known as Saptrishis★, who, after leaving their earthly existence, rose to the celestial sphere and now adorn the northern firmament, forming Ursa Major or the Great Bear which guides mariners on their way and so delights amateur astronomers and curious star-gazers.

★The Saptrishis: Atre, the one who gives protection from sin; Vashisht, the wealthy one; Kashyap, the one who protects his own body; the tortoise, Bhardwaj, the protector and benefactor of illegitimate children; Gautam, the disperser of darkness; Vishvamitra, the friend of the universe; Jamadagni, the one born of the sacrificial fire lit by the gods. The seven sages, after their earthly demise, ascended into the sky where they now display their glory in the constellation known as *ursa major* or the Great Bear.

When ripe in years, Gadhi, King of Kanyakunj, felt that he had savoured enough of the delights and pleasures of life, and decided to renounce his throne and seek the peace and piety of a secluded hermitage in the forest. He handed over the burden of royal responsibilities to his son, Vishvamitra, who was a worthy successor to him in every respect. Vishvamitra began to visit various parts of his realm, accompanied by a large army to inspire confidence in his subjects and to quell any rebellious elements as well as to safeguard the state from hostile encroachment. He hunted and made merry as he travelled, camping wherever took his fancy. One day he set up camp near the ashram of Vashisht who was away at the time of Vishvamitra's arrival. The King's soldiers began to cut down trees and bushes and to pluck the wild fruit and edible roots, of which there was an abundance in the area. When Vashisht returned he noticed the depredation and was greatly displeased, but when the King came to pay his regards, the sage welcomed him and served a most luxurious repast. The King was surprised to see the various dishes of meats and viands in this far away forest and asked his host how he had arranged the preparation of so many rich and delicious things.

'I have my Nandini,' answered the sage, 'and she instantly provides whatever I need.' Nandini, he explained, was a cow— the unrivalled kamdhenu, endowed with divine power, who was able to satisfy every conceivable wish of her master. To give greater conviction to his statement, Vashisht asked for more savouries and sweets, juice and liquor and fruit. These were immediately provided and heaped before the king. Then the cow brought rich clothes and costly jewels till King Vishvamitra was overwhelmed with astonishment, and an irrepressible desire to take Nandini away possessed him.

'Give her to me', he begged with joined hands, 'and I shall give a thousand of my best cows in exchange.'

Vashisht shook his head. He could not, would not, part with Nandini. The cow stood nearby and Vishvamitra gazed covetously at her thick neck, broad forehead, beautiful back, her long thick tail, broad horns and pointed ears, and teats large and swollen with milk. He said: 'Look, I shall give you ten million cows, all the wealth in my treasury, indeed my very throne and kingdom, if you hand me Nandini.'

But Vashisht was adamant; Nandini was not for sale or barter and he, therefore, could not satisfy Vishvamitra's desire. This put Vishvamitra in a temper. Burning with rage, he exclaimed, 'I am a Kshatriya and you are a Brahmin. Austerity, meditation and the study of scriptures is your duty. I am offering you a hundred million cows for Nandini. If you refuse, I shall, as behoves a Kshatriya, take her away by force.'

'You may do as you please', said Vashisht, 'for there is no need for you to consider the distinction between right and wrong.'

Vishvamitra ordered his men to tie the cow and lead it away. But the cow could not be moved. The soldiers used more force and beat her with whips and rods. But still Nandini remained immovable. She kept looking at her master and pleading for his help with loud moos. At last, Vashisht spoke to her, 'I see you and I hear your cries of pain. But I am a Brahmin, and it is my sacred duty to forgive and not use force. You may go with Vishvamitra if you wish. But I am not sending you away. The decision must be yours'. On hearing Vashisht speak in this manner, Nandini invoked all her spiritual powers, and procured enough men to defend her and push the King's soldiers away.

Vishvamitra was astonished. Overcome by a sense of frustration, he suddenly realized that a Kshatriya's ability to wield arms and command soldiers was nothing when compared to the spiritual might of a Brahmin, a might acquired by prayer, meditation and the practice of austerities. He therefore resolved to become a Brahmin, and giving up his throne, his kingdom and all his material possessions, he commenced a new way of life. He undertook a long and arduous course of austere living and meditation. Wild fruit and edible roots were henceforth his only nourishment. He fasted frequently. For a whole year, he subsisted on water alone, and then for several months on air alone. He suffered the heat and cold of the forest, and practised every kind of asceticism.

The gods were pleased with his selfless pursuit of piety and spiritual virtue and Brahma himself appeared before him and pronounced him a Brahmin. With this newly acquired status, Vishvamitra was able to wield power in all the three worlds. Sitting with Indra in his heavenly abode, he tasted somras, the liquor of the celestial beings. He was thus adequately equipped to deal with Vashisht.

It was not long before the opportunity presented itself. It so happened that King Kamlashpad, who was friendly with Vishvamitra and was seeking his support, happened to be out on a hunting expedition and wished to pay a visit to Vishvamitra after the conclusion of the hunt. As he was walking along a narrow forest path, he saw, coming towards him, Shakti, the eldest son of Vashisht. Shakti, in a gentle voice, asked the King to give way to him, as became a Kshatriya who must always defer to a Brahmin. The King refused to step aside or move back. Inevitably, the dispute became acrimonious. In a surge of temper, Kamlashpad struck Shakti with his whip. At this, Shakti, too, lost his temper and pronounced a curse, condemning the King to the status of a man-eating demon.

Vishvamitra had watched the entire scene from behind a tree. He now conjured up a demon, Kinkar by name, and prevailed upon him to enter Kamlashpad's body and devour Shakti for his first meal. This order was promptly obeyed by the demon-possessed King. Vishvamitra then urged the King to seek out and devour all the remaining ninety-nine sons of Vashisht one by one.

Seeing all his hundred sons fall prey to the demon-possessed King's insatiable appetite , Vashisht became so unhappy that he decided to end his life. He climbed up to the summit of Mount Meru and jumped off. But he sustained no injury. His long years of piety and prayer came to his help by softening the spot where he fell; so much so, that he thought he had fallen on a heap of soft cotton wool. Next, he walked into a blazing fire, but the flames did not touch him, and instead of burning him, gave him a pleasantly cool feeling. In sheer desperation, he now tied a large stone to himself and jumped into the heavy swell of the sea, but he was safely washed ashore by friendly waves. Dejected and defeated, he walked back to his hermitage.

The rivalry between Vashisht and Vishvamitra became increasingly bitter as time passed. Each strove to outdo the other in practising austerities and in striving to acquire more spiritual power. Vishvamitra, despite the acceptance of his elevation to the status of a Brahmin by all the gods, could not rid himself of the memory of his Kshatriya origins. On one occasion, he feared that Vashisht had surpassed him in the pursuit of divine power, so he planned to kill Vashisht. Vishvamitra knew that his rival was sitting on the high bank of the river Saraswati lost in meditation. So he called Saraswati, the goddess who ruled the river, and begged her to contrive to bring Vashisht before him. Knowing how powerful Vishvamitra was, the goddess was afraid to refuse this request. At the same time, she had great regard for Vashisht, and the thought of being an accessory to his murder daunted her. So she spoke to Vashisht and expressed her fears. Vashisht advised her to obey Vishvamitra's orders, otherwise he would pronounce a curse upon her. Greatly perplexed, Saraswati thought of a subterfuge. She wore away the base of the high bank on which Vashisht was sitting, thus making him fall into the river and be carried towards the spot where Vishvamitra was waiting for him. As soon as Vishvamitra saw his rival floating down the river towards him, he began to look for an appropriate weapon with which to strike a deadly blow to his enemy. But at this juncture, Saraswati, having carried out Vishvamitra's request suddenly turned the course of the stream, and Vashisht was carried back to safety and beyond his enemy's reach.

In his piety and determination to follow the path of righteousness at all times, Vishvamitra had a pragmatic, indeed, a truly Kshatriya way of solving life's problems. At one point of time the country was afflicted by a severe famine. For twelve long years the skies remained cloudless, and at night, there was no dew. Rivers, ponds and wells went dry, shops were forced to close, villages and towns became deserted. Death and starvation stalked the land. Old people were turned out of their homes by the young, where they starved to death out in the open. Cows, goats, sheep and buffaloes perished en masse. Even those creatures which survived had the appearance of the dead. There was no one to protect the weak and the needy. Men began to kill one another for food. In these conditions, Vishvamitra abandoned his home, his wife and child, and wandered the country in search of something to eat, for some sustenance to keep him alive.

In his wanderings he came to a village of Chandals, the lowest of outcastes and untouchables. Everywhere there were signs of uncleanness. In one house he saw broken pots, pieces of dog skin, the skulls and bones of pigs and donkeys, and the discarded clothes of men and women who had died. Nearby, cocks crowed and donkeys brayed and some Chandal men stood and shouted at each other nearby to where a pack of dogs sat.

Vishvamitra looked for meat to satisfy his hunger, but though he begged some of the Chandals to give him something to eat, they gave him nothing. Weak and exhausted, he fell down near the entrance to a Chandal house. Lying there, he continued to worry how he could obtain food. Searching the interior of the house with his eyes, he saw a piece of dog's flesh. At once, he began to devise a way of stealing the piece of meat, for there was no other way he could remain alive. In such circumstances, stealing would not be deemed a sin, for the scriptures have laid down that when it is a matter of life and death, a Brahmin is justified in stealing food from the house of a low-caste individual; failing this, he could rob a person of equal status and when an extreme emergency arises, he may steal even from one of higher status.

So, after a while, when the night was well advanced, he stealthily entered the house and laid his hands on the dog flesh. But he was seen and challenged by one of the family who was still awake. 'Who is stealing our meat?' the man called out. 'You will surely meet your end at my hands for this misdeed.'

'I am Vishvamitra', replied the sage. 'Hunger drove me to your house. If you are wise, don't kill me. A starving man is deprived of his sense of shame and is ready to steal. I have become very weak, and so I am no longer able to distinguish between sanctioned and forbidden meat. I begged for food in the village, but failed to get anything. Though I know that stealing is sinful, I thought it just and proper to steal in order to save my life.'

The Chandal replied, 'Respected Sir, do not act contrary to the dictates of dharma. It has been said that it is sinful to eat dog flesh, and even more so than the flesh of a jackal. Indeed, dog flesh is the worst and the most base food for a righteous person. Moreover, to steal from a low-caste Chandal is very very sinful. So, do not destroy the merit which you have so painfully acquired by your austerities and prayers.'

Vishvamitra insisted that he would be able to restore his saintly character by undertaking prayers and meditation but that he must first save his life by eating meat, and that to do something which keeps one alive must be free from guilt. He argued, 'Only when a man is alive can he practice justice and righteousness. Be persuaded that if I remain alive, I can by means of my spiritual strength safeguard what is right. Just as light dissolves darkness, so shall I be able to dissolve the consequences of all wrong and sinful deeds by austerity and knowledge.'

The Chandal replied that if he ate the dog flesh, Vishvamitra would not attain dharma, nor would the meat be a life-saving elixir to him. He repeatedly tried to dissuade Vishvamitra from carrying out his impious design. Vishvamitra, for his part, put forward pragmatic arguments to which he gave religious and sacred guise.

So, the two men continued to argue back and forth for a great part of the night, the Chandal accusing the sage of acting contrary to the dictates of dharma and the Vedas, while the sage strove to turn the argument away from him and to justify his conduct.

Finally, the Chandal by way of conclusion, said: 'Well, the wrong deed will bring punishment to the one who steals and not to the one from whom something is stolen.'

Vishvamitra then took the piece of flesh and, lighting a fire, cooked it, all the while chanting sacred mantras. He thus satisfied his hunger. Finally, he performed the appropriate ritual to neutralize the consequences of his doubly impious act.

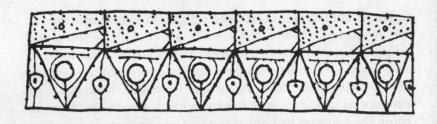

Nala and Damyanti

In the course of the thirteenth year of their exile, Bhim and Yudhisthir were lamenting their plight. Bhim repeatedly complained of Yudhisthir's unholy passion for gambling which brought about the loss of their kingdom and all their wordly possessions. 'Ah, woe is me,' Bhim exclaimed, 'there is no one more miserable than I.'

The sage Vridhasva happened to arrive at this moment, and greeted the Pandavs. 'I shall relate to you,' he said after a brief rest, 'the story of a king who was even more miserable than you, and whose suffering also resulted from his passion for gambling.' Vidhasva then related the story of Nala and Damyanti.

In the land of Nishada, there was a King named Nala who was brave, handsome, virtuous, and skilled in the arts of war and the management of horses. He was magnanimous and open-hearted like Manu, but he had a passion for gambling. In the brilliance of his person, he equalled Surya, and the nobility of his thought and deed was no less than that of Indra, the lord of celestial beings. It was but natural that the fame of his personal charm and regal attributes should spread to distant regions and reach the ears of princess Damyanti, daughter of King Bhim of Vidarbh. The young princess, too, was unequalled in beauty, charm and feminine graces.

When the tales of Damyanti's royal attributes reached Nala, he soon became enamoured of Damyanti, and she reciprocated Nala's sentiments. Thus, these two young persons of two distant noble families found themselves deeply in love, without having met or seen each other.

One day, Nala was walking in his garden and thinking of Damyanti, when he saw a flight of golden-winged swans moving about the flower-beds. On a sudden impulse, he chased and caught one of them. The swan pleaded that he would do anything the King wished if he were set free. Indeed, he would go to Damyanti and spell out to her all the virtues possessed by Nala. This would assuredly make her fall passionately in love with him. Nala agreed to the proposal.

As soon as Nala released the swan, the golden bird and its companions flew off in the direction of Vidarbh. The swans soon arrived in the royal garden where Damyanti and her companions were playing games. The girls began chasing the swans. Each time Damyanti came near a swan, the bird spoke in a human voice, saying, 'In Nishad there is a King named Nala who is handsome like the Ashwins and is unmatched by anyone on earth. If you marry him, your purpose of life will be fulfilled. Both of you are worthy of one another.'

Damyanti, for her part, told the swan to go back to Nala and tell him she felt the same way towards him. So the swan returned to Nishad and delivered Damyanti's message to Nala.

Damyanti now began to suffer from the pangs of love. Each day that passed made her more melancholy and restless. She kept looking up at the sky with an abstracted expression and lost interest in food and drink and the daily routine of life in her father's palace, as the pain inflicted by Kama, the god of love, made her morose and uncommunicative. Her companions guessed the true cause of her unhappy state and related to the King all that had happened. Bhim considered the matter and decided to hold a swayamvar ceremony, so that his daughter could choose a husband worthy of her. He accordingly sent out invitations to all the kings of the realm to come and take part in the ritual. The gods in their celestial abode also came to know of what was going on in Bhim's court. Hearing that a young princess of unparalleled beauty was about to choose a spouse for herself, they at once decided to offer themselves to Damyanti.

Mounting their heavenly chariots, they started out for Bhim's palace. But, as they proceeded on their way, they saw Nala going in the same direction. Seeing his extraordinary handsomeness and dignity, they began to think that despite their godly attributes, they fell short of Nala's merit and could scarcely hope to compete with him at the swayamvar. So they left their chariots, and descending to earth, accosted Nala, saying to him, 'Royal Sir, you are a truth-loving person. Please be our messenger and help us.'

Nala, without giving the matter a thought, acceded to the request of the gods, even before asking who they were and to whom the message was to be carried. But when the gods revealed their respective identities as Indra, Agni, Varun and Yama, and when they further asked him to go to Damyanti and tell her that she should choose one of the four gods as her husband, he was shocked and began to excuse himself by saying that he himself was on his way to the swayamvar with the set purpose of winning Damyanti as his wife, so how could he plead the cause of anyone else? But the gods insisted on binding him to the undertaking he had already given. 'You have', they protested, 'already promised to be our messenger. How can you break your word now'. When Nala argued that his passage into Damyanti's apartment would be barred by guards, the gods assured him that they would contrive his free entry to the princess's presence by using their divine spiritual powers.

And so, Nala submitted himself to what was no less than a divine command. As promised by the gods, he walked past the guards without them seeing him and entered the chamber where Damyanti was sitting with her companions. He felt that the slender-waisted, bright eyed, beautiful young princess surpassed even the glory of the full moon. One look at her made Nala the irredeemable victim of Kama deva's arrows. But he maintained his poise and remained outwardly unruffled. 'I am Nala, King of Nishada', he said, and I have come to you as the messenger of the four gods, Indra, Agni, Varun and Yama. Each one of them is desirous of winning you as his consort. You may choose the one whom you wish to be your lord.'

Damyanti bowed her head in respectful obeisance to the invisible gods. Then she smiled and said, 'O King, I have already dedicated my mind, my life and all I possess to you. Grant me your love and do what you deem just and proper. Ever since I heard the swan with the golden wings speak of you, I have been suffering in mind and body. It was to win you that a swayamvar was arranged, and I have vowed to have you for my husband. If you abandon me, I shall immolate myself by entering a consuming fire, drown myself in deep water, take deadly poison, or hang myself.'

Nala tried to plead, once again, the cause of the celestial suitors, because he felt compelled to keep his promise and to escape the dire consequences of refusal. Damyanti, however, was firm. With her eyes streaming with tears and her countenance a picture of deep sorrow, she replied, 'O Lord of your earthly realm, I bow to my heavenly suitors, but I will choose you and only you for my husband at the swayamvar.'

With his heart beating wildly with joy, Nala exclaimed, 'How can I forswear my vow and break my word given to the gods? But if you can find a way out of this impediment, I shall, with all my heart, comply with your wishes and take you for my beloved spouse.'

'Well then,' said Damyanti, 'come to the swayamvar and I shall choose you in the presence of the gods. No blame will thus attach to you.'

Nala went back to where the gods were waiting for him and narrated to them the whole incident, both honestly and truthfully.

The swayamvar hall was exquisitely furnished in truly regal style. Tall golden pillars supported the high portal arch towering above the richly decorated floor, and the seats of the assembly hall and all the decoration proclaimed the wealth and importance of King Bhim. At the auspicious moment, which had been earlier determined by calculating the favourable position of the moon and the planets, the competing kings and princes came and took the seats assigned to them. In their rich regalia and their resplendent crowns, they made a most impressive gathering, though they were possessed by a mixture of hope and anxiety.

Damyanti entered, holding in her hand the swayamvar garland. All eyes at once turned to her and stayed on her dazzling beauty. The court officer began introducing the suitors, naming them one by one and proclaiming their personal attributes.

Damyanti saw that there were now five Nalas among the suitors, all looking exactly alike and in no way distinguishable one from another. She guessed at once that the four divine suitors mentioned by Nala had resorted to this clever strategy to defeat Nala's own purpose. She pondered over the matter, paused in front of the five Nalas, and offered a silent prayer to the supreme divinity to disenchant her vision so that she could choose the true, mortal Nala. Then she again looked at the five Nalas more carefully, this time to discover any special distinguishing sign indicative of the real Nala. She now observed several differences. Only one out of the five blinked his eyes, had drops of perspiration on the forehead, and had his feet firmly planted on the ground like all mortals do. The remaining four stared in front of them without blinking their eyes, had no trace of perspiration on their foreheads, and their feet did not quite rest on the ground.

Damyanti therefore stepped forward and shyly bestowed the garland on the mortal Nala, the true object of her love. Nala, finding himself openly preferred to the four celestial suitors and all the other kings and princes present in the hall, vowed to love Damyanti till his dying day and promised to be a devoted and faithful husband to her.

The gods accepted their defeat without rancour, and blessing the bridal pair, commended them to a happy, virtuous and long married life. The wedding ceremony was then performed, and in due course, the royal guests left for their respective domains. After a few days' stay in Bhim's palace, Nala and Damyanti, too, left for Nishada.

There they spent many happy years of married life, Damyanti giving birth in due course to a son and a daughter, who were named respectively, Indra Sen and Indra Sena.

All this while, two wicked individuals, Dwarpayug and Kalyug, who lived in the upper world and claimed status in the heirarchy of celestial immortals, were plotting to disrupt the even tenor of the Vishada monarchy and to destroy the happiness of the royal couple. These two immortals had also wished to take part in Damyanti's swayamvar to compete for her favours. But they had not arrived in time, but had met Indra, Agni, Varun, and Yama on their return to their celestial abode after the wedding ceremony.

Indra told Dwarpayug and Kalyug that Damyanti had chosen Nala as her spouse, adding that it was well that this should be so, for Nala was endowed with all the human virtues and had always trodden the path prescribed by dharma. Kalyug was so incensed by this news that he gave vent to an angry outburst: 'As she has chosen a mere mortal over heavenly suitors, she must pay a dire penalty.'

Indra tried to argue with Kalyug and pacify him but Kalyug remained adamant, saying to his companion Dwarpayug, 'I cannot contain my anger. Whatever happens I shall enter Nala's body, and with my divine powers, deprive him of his kingdom. Dwarpa, help me by entering his gambling pieces and ruin him when he starts playing.'

After forming this unholy pact, Kalyug and Dwarpayug entered into Nala's kingdom incognito and took up residence near Nala's palace. Kalyug constantly kept watch over Nala's activities, waiting for an opportunity for Nala to deviate from the path of virtue: perhaps some inadvertent act or omission, resulting in a minor impurity would furnish him the chance to enter Nala's person so as to achieve his purpose. But Nala was always punctilious in the strict observance of sacred rites and virtuous conduct.

Thus passed twelve years of peace and happiness for the royal couple. Then, one evening, Nala forgot to wash his hands after urinating. Kalyug, at once seized his opportunity and entered Nala's body. Simultaneously, he began to urge Nala's brother, Pushkar, to play a game of dice with Nala, suggesting to Pushkar that by this means, he could win Nala's kingdom and all his wordly possessions. Thus impelled by greed and ambition, Pushkar repeatedly invited his brother to a game of chance. Nala finally succumbed to his innate passion for gambling which had lain dormant over the years, and sat down to play dice with his brother. The dice had been heavily loaded by Dwarpayug's cunning and within a short time, Nala lost all his gold and silver, and his chariots, charioteers, and all their equipment. He paid no heed to the protestations of his wife, and turned a deaf ear to the warnings of his ministers and the agitation of his subjects. So, in the course of a few months, he lost his kingdom and all he possessed.

Damyanti foresaw a bleak future for herself and her children, and so she sent for Nala's foremost charioteer and sent her son and daughter with him to live with her father, Bhim, in Vidarbh. On Damyanti's advice, the charioteer entered the service of King Rituparn of Ayodhaya. Nala and Damyanti, deprived of all they possessed, left their royal home, and each clad in a single piece of cloth, took voluntary exile.

Thus began a period of extreme adversity for the royal couple. For three days and nights, they went without food. Nala, with Damyanti following behind, then set out for the forest to look for whatever sustenance the wild trees and bushes could provide.

After some days Nala saw a flight of birds. Hoping to capture them, he threw the single piece of cloth that formed his clothing over them. The birds immediately rose up as one and soared away out of sight taking Nala's clothing with them. Nala, thus deprived of the last material possession, suggested to Damyanti that she should seek refuge in her father's home. He pointed out the way which would take her to Vidarbh after traversing a not too difficult route. But Damyanti replied that she would not leave her husband. If, however, Nala went with her, she was willing to go to Vidarbh.

Nala, however, did not relish the prospect of going to Bhim's palace. 'How can I,' he exclaimed, 'go to Vidarbh in my present state? Previously, my splendour and majesty were no less than Bhim's. When I entered his presence, it gave you immense joy and you felt so proud of me. Now my appearance shocks and mortifies you.'

So they stayed with each other, Nala sharing Damyanti's sari by taking half of it to cover his own person. Thus they proceeded through the forest together.

After some time, Nala and Damyanti both felt tired, and coming to a secluded, sheltered spot, they sat down to rest. Damyanti presently went to sleep, but Nala's mind was tortured by their wretched plight. He was confused and could not think clearly. What should he do, and how should he relieve the suffering of his loving and devoted wife? Finally, telling himself that if she were left alone, she would in all certainty make her way to her father's protection and a home that could provide her comfort, he cast off the half of the sari which covered him and moved away, leaving her lying asleep on the ground. Looking back for a moment as he walked away, he remembered her glowing beauty and her queenly, alluring gait of former times, and was overcome by remorse and anguish; feeling as if his heart was breaking, his eyes flooded with tears. It was thus with great reluctance that the grief-striken Nala left his wife and walked away into the forest.

When Damyanti awoke and found that Nala was not by her side, she started weeping and moaning. Damyanti called out to Nala across the forest in a voice heavy with anguish, reminding him loudly of his vows and his duty to remain with her and to protect her. Wailing and weeping, she roamed the forest, always hoping to find Nala behind the next tree, or to see him coming towards her out of the distance. She stumbled on, over rough and thorny paths, meeting predatory wild animals prowling on their rounds. She once escaped a boa constrictor who seemed to be waiting for her, and then repelled the advances of the hunter who killed the boa and who, finding her alone and unprotected, tried to molest her.

After some days of aimless wandering, she chanced upon a caravan of merchants. Seeing her helpless condition, they comforted her and gave her sustenance and shelter. She told the story of how she came to be in such a perilous state and related the harrowing experiences she had gone through. Sadly, the merchants told her that they had not seen Nala anywhere on their route, but that they were on their way to see Suvahu, the king of the Chedis, and she was welcome to go with them.

After a few days' journey, the caravan halted at the banks of a lake fragrant with lotus flowers and sheltered by a thick forest all around. Green trees hung laden with wild flowers and edible fruit. A bed of soft green grass extended from the lake, providing an excellent campsite. The merchants, tired after the day's journey, lay down to rest and were soon fast asleep.

Suddenly, the peace of the night was shattered by an invading herd of wild elephants which had been attracted to the camp by the merchants' tame elephants. The animals trampled through the camp, destroying tents and bags full of merchandise, even crushing underfoot unsuspecting merchants. There ensued complete chaos, with men and women running hither and thither. A number of pack horses and camels were killed or maimed. Those who survived the tragedy began weeping and moaning, bewailing their losses. One merchant then said that their ill-fortune was all due to the bad luck brought by the wretched destitute woman who had joined the caravan a few days earlier. This was soon taken up as a refrain, everyone denouncing Damyanti, calling her an evil witch, a shameless woman who wore only a tattered piece of cloth which hardly covered her person. 'Let us kill her,' they shouted. 'Stone her to death and rid us of the evil presence, or she will bring further misfortune and losses to us all.'

Damyanti, lying petrified with fright, heard these words and decided to run away. Taking advantage of the darkness of the night, she quietly slipped away into the forest and ran off. She continued her progress, all the while bewailing her ill-fortune till the evening of the following day. She was now approaching the capital city of the Chedis where King Suvahu resided. Clad only in her half garment, she quietly entered the city, her face and limbs covered with dust, and her hair dishevelled after her long tramp, giving her the appearance of a mad woman. A band of curious urchins began to follow her. It was thus that she arrived in front of the royal palace.

The Queen Mother saw her from the terrace of her apartments. She thought that, despite her pitiable state, Damyanti seemed to possess remarkably good looks. Out of pity for the wretched woman, the Queen Mother sent her nursemaid to fetch her. When Damyanti stood before the Queen Mother a little later, the Queen Mother was astonished by the real beauty that lay beneath all the dust and misery, and by Damyanti's regal bearing. On being questioned, Damyanti related the entire story of her misfortune, saying that she was seeking her husband who had left her alone in the forest.

The Queen Mother comforted the distressed Damyanti, and asked her to stay in the palace as her personal companion. Meanwhile, royal messengers would be sent to look for Nala and bring him to her. Damyanti replied, 'I am willing to stay, but I have made certain vows which must be respected: I cannot eat the left-overs of another; I shall not wash anyone's feet; I shall not converse with a strange man, and if anyone tries to tease or molest me, you must adequately punish him. And may I be allowed to give instructions to your emissaries when they go to look for my husband.'

The Queen Mother readily agreed to respect Damyanti's wishes. She then called her young daughter, Sunanda, and asked her to show Damyanti to Sunanda's apartments and to take her as her companion.

In the meantime, in the course of his journey through the forest, Nala saw a great fire raging in a clump of trees, and, hearing a voice calling out, 'Nala, Nala, come and save me; hurry, O pious King,' ran up to the burning trees where he saw a gigantic serpent sitting curled up in the circle of fire. 'Save me', cried the serpent, 'I am Karkatok, the most powerful denizen of the serpent world, and I am in this state because of a curse pronounced by the sage Narad, who I tried to deceive. As a consequence of that curse, I am obliged to stay here till your arrival. O Nala, please lift me up out of this dangerous fire. I shall be your friend and help you in every way.'

Nala thereupon lifted up the serpent, which had reduced its weight by its innate power, and carried it to a safe distance from the fire. 'Don't put me down, yet,' pleaded the serpent. 'Walk for some distance, holding me up and counting your steps. I shall confer a boon on you.'

Nala did as he was requested. He walked slowly, counting loudly, 'One, two, three. . .' When he came to the figure ten and said '*dash*', the serpent bit him, for dash means bite as well as ten. To Nala's utter surprise, his appearance was immediately changed by the bite, while the serpent assumed his original form and said, 'I have changed your appearance so that no one should recognize you. And the poison I have injected into your body will not harm you, but it will torture Kalyug who has been sitting inside you and making you suffer in various ways. Henceforth, you will be safe and immune to physical injury or any curse. Go to Ayodhaya and enter the service of King Rituparn. Tell him you are Bahuk, an expert in the training and handling of horses. Rituparn is a highly skilled player of the game of dice. In return for your services he will teach you how to win in this game. Take the clothes which I will give you. When the appropriate time comes, pronounce my name 'Karkatok', and put on these clothes. You will immediately resume your original appearance.'

Having said this, Karkatok disappeared. Obediently Nala followed the serpent's instructions, and going to Ayodhaya, offered his services to King Rituparn, saying that he was an expert in the management and training of horses and that he was also a master of the culinary arts, besides being an experienced financial accountant. Nala, now Bahuk, also claimed the ability to handle many other jobs. Rituparn at once employed him and agreed to pay him a monthly salary of ten thousand gold sovereigns. He also provided two assistants, Vershneya and Jival to help him in discharging his duties.

And so began Nala's secure and carefree life in Ayodhaya. But his heart was torn by memories of Damyanti and his deep anxiety about her whereabouts and her welfare. He kept constantly repeating to himself, 'Alas! How is she faring; is she hungry and thirsty; who is looking after her?'

One night, as Nala was uttering such things, Jival asked him, 'Bahuk, who is the woman for whom you are so worried?' Nala replied, 'Jival, she is the virtuous wife of a stupid man who left her for some reason, and now he spends sleepless nights, yearning for her presence.'

Bhim, in the meantime, had sent out his emissaries to look far and wide for Nala and Damyanti. He had no knowledge of their whereabouts; the need for parental love and care for his grandchildren made it all the more imperative that he should find Nala and Damyanti. So, he sent off Brahmins to look throughout the country, charging them to make a diligent search for his daughter and son-in-law and promising a rich reward to whoever should bring back either of them or discover their whereabouts.

The search would, of course, have proved fruitless, if it were not for one of the Brahmins, named Sudev, who, in the course of his search reached the capital of the Chedis, and who, in King Rituparn's palace, came upon Princess Sunanda's companion, who he found bore a most striking resemblance to the young princess Damyanti as he remembered her from the King Bhim's palace some years previously. When he questioned her closely, Sudev was convinced beyond doubt that she was indeed Damyanti. And so the secret was out, and the identity of the handmaiden became known to the Queen Mother and the King of the Chedis. The Queen Mother said that good fortune had indeed come to her, for the Queen Mother was the sister of King Bhim's consort, Damyanti's mother: the Queen Mother was thus the young handmaiden's aunt. So the Queen Mother would be happy beyond measure to send her beloved and virtuous niece to her parents in Vidarbh.

Preparations for Damyanti's departure were immediately taken in hand. And so, after four years of sorrowful absence, Damyanti was happily reunited with her parents and her young children. The Brahmin Sudev was appropriately rewarded with land, riches and a thousand cows.

Bhim now intensified the search for Nala. He admonished his men to spare no pains in tracing Nala's whereabouts and to bring him back to his wife and children. The Brahmins, before setting forth on their quest, went to Damyanti and asked her if she could give them any clue or any message which would help them in their undertaking. Damyanti said to them, that wherever they were to go and to whomsoever they met, the Brahmins should say, 'Beloved gambler, where did you go after tearing off a portion of your loving wife's only garment, leaving her fast asleep!'

Thus briefed, the Brahmins travelled far and wide in all directions, crossing streams and rivers, visiting towns and the countryside, repeating the question framed by Damyanti to all and sundry. For a long time, their labours remained unrewarded. Then at last, one of the Brahmins, Parnad, on returning to Vidarbh went to Damyanti and told her that when he was making enquiries in the capital city of Ayodhaya, an ugly man came up to him and said that he was Bahuk, that he was employed by King Rituparn as a charioteer, and that he was also an expert in horse management and cooking. Bahuk, with his eyes full of tears and sighing deeply, had said, 'A chaste and devoted woman should not be angry with one whose only garment was carried away by birds, and who leaves his sleeping wife to search for food and drink.'

Parnad's report convinced Damyanti that Bahuk was none other than her royal consort, Nala, and that she would soon be able to see him again. So she called Sudev, who had been successful in finding her, and in her mother's presence she instructed Sudev to go at once to Ayodhaya and to tell King Rituparn that the Princess Damyanti was holding another swayamvar ceremony to choose a husband because she thought that Nala was no longer alive. The swayamvar would take place within twenty-four hours. Sudev travelled at great speed and delivered Damyanti's message to Rituparn.

Rituparn was so thrilled and excited by the prospect of being chosen by Damyanti that he immediately summoned his charioteer, Bahuk, and told him that Damyanti was to hold a swayamvar on the following day, and that he had decided to be present at it. He ordered Bahuk to make haste and to make his chariot ready with a team of the fastest horses from the royal stables, so that he should arrive at the capital of Vidarbh in time.

Bahuk went to the stables and carefully selected four lean horses who were, in his view, the fastest and the most powerful steeds. With these he made the chariot ready. Rituparn expressed his fears about the competence of such lean horses to perform the formidable task to be undertaken by him, but Bahuk allayed the King's fears and pointed out that the horses chosen by him were, in fact, the strongest and the fastest, their wide nostrils and mouths, the ten-hair tufts on their foreheads, and their strong flanks, legs and back, all being indications of their superior breed and unrivalled performance. Bahuk assured the King that they would take him to Vidarbh well in time for the groom-choosing ceremony.

Bahuk was proved right, and Rituparn arrived at Bhim's palace before sundown of the day before that announced for the swayamvar ceremony.

Rituparn was surprised to find that he was the only suitor for Damyanti's hand. And though he saw no apparent arrangements for the ceremony, he nevertheless decided to call on his host and exchange greetings with him. Bhim, for his part, could not understand why his royal visitor had taken the trouble to come all the way from his own country to pay a courtesy call, and wondered why King Rituparn had ignored so many other rulers whom he had passed in the course of his journey, choosing only him for the honour. However, the two Kings maintained truly regal discretion and exchanged friendly greetings, talking of various matters. King Bhim then made adequate arrangements for King Rituparn's stay and entertainment for as long as he chose to partake of this hospitality.

From the terrace of her apartments, Damyanti had seen Rituparn's chariot arrive. She had watched the horses being unyoked by Bahuk, whose dwarfish appearance bore no resemblance to her tall handsome husband. She was torn between hope and apprehension. The information brought by Parnad and the peculiar sound of the chariot wheels when Rituparn's procession had arrived in the palace courtyard had led her to hope that she was, at long last, going to see her lord, disguised as a charioteer. But now that she saw the charioteer unharnessing the horses, she began to fear that fate was deceiving her once again. She summoned one of her maid servants and asked her to go immediately and speak to the charioteer. 'Ask him who he is, and repeat to him the question I had instructed Parnad to ask, and bring me back his answer.'

The maid servant accordingly went and spoke to Bahuk. She then returned to Damyanti and reported, 'He says that a chaste and devoted wife should not be angry with one whose garment has been carried away by birds, and who leaves his sleeping wife to look for food and drink.' The maid servant added, that as Bahuk made this answer, he broke down in tears.

Damyanti's fears now vanished, but she wanted to be certain beyond all shadow of doubt. So, she again sent the maid servant to go and secretly watch every word and action and every movement of the charioteer and to bring back a full report of what she observed. After some time, the maid servant brought a portion of the meat dish which Bahuk had cooked for Rituparn. No sooner had Damyanti tasted the meat than she recognized Nala's skilful hand in cooking, the dish being made so perfectly with such a delicate taste. By way of a final test, Damyanti asked the maid servant to take her children to Bahuk and let him see them.

Bahuk, on seeing his son and daughter after years of separation, could not conceal his happiness and shed tears of joy. He exclaimed that the children were just like his own whom he had not seen for many years. Now, being completely convinced of Bahuk's real identity, Damyanti called her mother and told her all that she had done and heard. The Queen went to Bhim and related the entire story to him. Bhim immediately gave his consent to Damyanti calling Bahuk to her apartment. The moment Bahuk saw Damyanti, he was overwhelmed with a strange mixture of joy and grief for his wife: his royal consort was standing before him clad only in the half-garment which she was wearing when he had left her sleeping in the forest. Damyanti fell into his arms, and sobbing with happiness, clung to him.

There followed an exchange of loving recriminations about each other's errors and omissions and the unhappy events ordained by the angry gods. Soon, joy and contentment drove out all thought of their long and patient suffering. Nala put on the two pieces of cloth given him by the serpent Karkatok and pronounced Karkatok's name. His beauty of person and his valour were then restored to him.

Nala had in the meantime received full instruction in the game of dice. So, he went back to Nishada and challenged his brother Pushkar to play again. Pushkar, thinking that he would win as before and that he would be able to take away whatever remained of Nalas' possessions, accepted the invitation. But now, with his newly acquired skill, Nala won back all he had lost: his treasures, his wealth, his kingdom and the throne on which he previously used to sit. But moved by fraternal regard and compassion, he forgave Pushkar and gave him land and houses and enough money to live in comfort for the rest of his life.

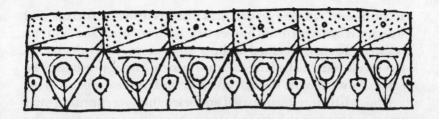

A Woman's Revenge

Bhishm, the Terrible One, ancestor of the warring Pandavs and Kauravs, had forsworn the right to sit on the throne, but continued to act as the regent of his younger brother, Vichitravir. In course of time, Vichitravir attained manhood and Bhishm thought of finding a suitable spouse for him.

The news spread abroad that the King of Kashi was to hold a swayamvar ceremony for his three grown-up, beautiful daughters. Bhishm accordingly took a chariot and drove to Kashi, where he found that the husband-choosing ceremony for the three brides-to-be was already in progress.

A large number of kings and princes had come, hoping to be fortunate enough to be chosen by one of the young princesses. The King's three daughters, Amba, Ambika and Ambalika were sitting in the places appointed by the King and the virtues and exploits of the suitors were being proclaimed by the royal herald. Bhishm's arrival set tongues awagging. One man said the old man still looked full of fire, another remarked that his vow of celibacy was a piece of hypocrisy, else why, in his old age, had he come to attend a swayamvar. Bhishm took umbrage at these murmurings, and standing up, declared that he was taking away all three young princesses. He then took them by force and pushed them into his chariot, telling the astonished assembly that he was taking recourse to the seventh recognized form of marriage, namely the Rakshasa or demon form, according to which a Kshatriya takes his bride by force.* The warrior who overcomes and destroys his enemies and wins his bride by force is deemed to be a most virtuous bridegroom. He threw out a challenge: 'I am taking these princesses by force. Any one who dares, can come and oppose me.'

Thus challenged, the royal suitors, who were fuming with anger, rose up in a body to attack Bhishm so as to rescue his fair capt'ves. Single-handedly, Bhishm defended himself. Braving their arrows and lances, he inflicted grievous wounds on his assailants, disabling them one by one, until they were forced to abandon their task and to let Bhishm take the princesses away.

However, King Shalv, who had hoped to be garlanded by the eldest princess, Amba, gave chase to Bhishm. King Shalv had a special grievance against the Terrible One because he already had an understanding with Amba, who had made up her mind to take Shalv as her life partner. Seeing Shalv following him, Bhishm began to burn with rage. Fear was something he had never known, and he had always been an indomitable defender of the Kshatriyas. He turned his chariot round and returned to deal with Shalv. The other rulers now stood to one side to watch the conflict.

The two warriors rushed at one another like two fighting bulls, Shalv firing a salvo of arrows at Bhishm. The royal spectators cried 'Bravo, well done.' Bhishm then countered by hurling his Varun missile which killed all four horses of Shalv's chariot. Shalv changed chariots and renewed his attack. Bhishm, countered by launching the Indra weapon, again slaughtering all four of Shalv's horses. He then leapt from his chariot and seized hold of Shalv, but seeing him totally defeated and helpless, released his hold and spared the young king's life. Shalv accepted his defeat and went back to his state, resigning himself to the peaceful administration of his realm.

* Eight types of Hindu Marriage:
1. Brahma, 2. Daiva, 3. Arsha, 4. Priyajapatya, 5. Asur, 6. Gandharv, 7. Rakshas, 8. Paishach. The first four types have been laid down as suitable for Brahmins, the first six for Kshatriyas, but for a Kshatriya King even the seventh type is permissible.

Following his victory, Bhishm conducted the princesses to Hastinapur where he presented them before his step-mother, Satyavati. She immediately gave her consent to all three being married to Vichitravir. On hearing this, Amba, the eldest sister spoke up. Addressing Bhishm she said, 'You are a righteous man, that is why I am making this submission to you. Before the swayamvar was announced by my father, I had pledged myself to King Shalv of Kashi. My father had knowledge of this and had resolved to give my hand to Shalv. So, I consider only him to be my husband. You have been born in a family of upright and virtuous people. It would not be consistent with your family honour to keep me here. King Shalv will be waiting for me. Send me to him and grant my prayer because I have heard that you are the soul of piety and virtue.'

Bhishm readily agreed, and made arrangements for Amba to be conducted to Kashi. He deputed Amba's nursemaid and a group of aged Brahmins to accompany her and look after her needs on her journey. In due course, she reached Kashi and standing before Shalv, said, 'I have come to you to fulfil my vow to have you and no one else as my husband. Accept me as your bride and wife.'

Shalv smiled and replied, 'Beautiful one, you have already visited another's home, so I cannot marry you. Go back to Bhishm. He took hold of your hand and made you sit in his chariot. You did not object. Moreover, Bhishm won you by the force of arms, so I cannot let you remain in my house. How can a wise and true follower of the sacred precepts who is always admonishing others to tread the narrow path of virtue, do otherwise? Go back to Bhishm or anywhere else you choose to go— but don't delay.'

To this, the love-stricken Amba replied, 'Slayer of foes, do not speak like this. Neither when Bhishm took me by force, nor at any other time, have I entertained any feelings of tenderness toward him. When he drove me away, after defeating all the kings who came to my rescue, I was weeping. He took me away by force of conquest: I cannot be blamed for that. I love no one but you. It is wrong to turn away a woman who is blameless and who has come to seek refuge. I asked Bhishm permission to come to you, which he granted. Bhishm had wanted brides for his younger brother; he did not wish to marry me himself. I swear solemnly that I love no one but you, that is why I have come to you. Accept me as your wife.'

But Shalv persisted in rejecting her even after repeated pleadings. Finally, shedding tears of anger and mortification, Amba retorted, 'Very well, you renounce me. Be it so. I have spoken nothing but truth, and I shall get protection wherever I go.'

'Go, go to Bhishm,' Shalv retorted. 'He is the Terrible One, and I do not wish to live in perpetual dread of him.'

Amba, disappointed, crestfallen, and heart-broken, left Shalv's town. Once outside, she began to debate with herself. 'Woe is me,' she moaned. 'No woman on earth can be more ill-fated than I. No one can be suffering more than I. Deprived of my relatives, and rejected by Shalv. I cannot go back to Hastinapur and face Bhishm. Am I the one who is at fault, should I put the blame on Bhishm, or should I blame my father for the folly of holding a swayamvar? No, I fear it is I who must bear the most blame. When Bhishm was fighting with all the kings who were trying to rescue me, I should have jumped down from the chariot and should have gone to Shalv. My present wretched state is the fruit of that inaction. Cursed be Bhishm and cursed be Shalv and cursed be my father, but I, too, am accursed, for so is my fate. But they are to blame for my miserable plight. True, a man reaps what he sows, but there is no doubt that the root cause of this injustice is Bhishm. So, I must take my revenge for what he has done to me, whether I have to fight him or whether I achieve my aim by performing austerities and prayers. First, I must find out who can overcome Bhishm in a fight.'

Thus, turning over the conflicting emotions of her heart, Amba walked on till she chanced upon a hermitage of holy men. She related to them the entire tale of her misfortune, with deep sorrowful sighs and frequent shedding of tears. When she had finished, they said that it was beyond their power to relieve her suffering. Amba cried out, 'Have pity on me. I am ready to undergo the severest austerities to achieve my purpose. I must be suffering because of some misdeeds in my past life. I have no wish to go back to my father or any other relative. I am determined to undertake any penance to acquire spiritual power. Please, let me stay in this hermitage and execute my design.'

The sages debated the matter among themselves. One of them suggested that Amba should be sent back to her father in Kashi. Another said that she should go to Bhishm and point out his error to him and make him see sense. Another was of the opinion that Shalv should be asked to relent. Yet another pointed out that Shalv had already said a categoric no to Amba, and it was pointless to approach him again. Finally, they all agreed that they really could not do anything to help Amba achieve her desire.

'You should abandon the idea of undergoing life-long austerities, and go back to your father. He alone is capable of alleviating your suffering. With him you will live in comfort and happiness. Listen, you are a young and attractive woman, and no one can look after a girl better than her father. A woman's saviour is either her father or her husband and, in adversity, it is the father to whom the woman turns. You are a princess, brought up in luxury with tender care, you will not be able to bear the hardships of an austere life. And your stay here with men will set tongues wagging. You will be spared all this in your father's home.' One sage even added, 'If you live in this forest hermitage, it may so happen that a king or a young prince, coming to hunt, may chance to see you. He may want to take you away. So, just abandon the thought of living a life of abnegation and austerity.'

But Amba was adamant. She argued that in her father's home she would inevitably invite the calumny of everyone. She was determined to remain in the protective security of the hermitage and to undertake the severest austerities to attain her purpose.

At this juncture, Hotravahan, a royal rishi deemed the wisest and most learned sage, happened to visit the hermitage. He was Amba's maternal grandfather, and on hearing the distressing tale of her misfortune, he comforted her and declared that she should on no account go back to her father. With deep love

and concern he told her, 'I am your mother's father. Have faith
in me. I advise you to go to Parshuram. He will relieve your
suffering by killing Bhishm. In this manner your wrong will
be righted.'

Amba bowed her head in consent, and with her eyes full of
tears, said to him, 'Be it so. But grandfather, how can I find
Parshuram?'

Before Hotravahan could answer Amba's query, Akritvran
arrived. He was also a renowned sage and a personal friend of
Parshuram. The entire matter was related to him in detail. He
said Parshuram would be coming to the hermitage the next
morning, and he would, no doubt, solve Amba's problem. Amba,
on hearing this, said, 'Holy one, the fear of condemnation and
my own sense of shame forbid me to go back to my father, but
I shall do what Parshuram says and accept his decision as a
sacred precept.'

Akritvran replied, 'There are two courses open to you. If
you wish to make Shalv accept you as his bride, Parshuram
will, without a doubt, prevail upon Shalv to do so. If, on the
other hand, you wish to see Bhishm defeated in combat by
Parshuram in order to take your revenge, the great Brahmin
warrior will gratify that desire too.'

'Holy one,' Amba replied, 'When Bhishm carried me away,
he did not know that I had given my hand in marriage to Shalv.
He acted in ignorance. So, let the matter be adjudicated by you
as to whether Shalv should be compelled to marry me or whether
Bhishm should be put to shame and slain. Whoever is responsible
for my pain and suffering should receive his just punishment.'

The rest of the day and the following night passed in the dis-
cussion of Amba's problem in all its detail. All possible solutions
were considered and reconsidered without any decision being
taken. In the morning, the great Parshuram accompanied by
his retinue of disciples arrived at the hermitage. He was greeted
with great respect and offered refreshment. Hotravahan then
introduced his granddaughter and asked Parshuram to relieve
her suffering. Amba placed her head to Parshuram's feet in
submission and respect and stood before him, shedding tears of

sorrow. The sage observed her incomparable beauty and gently asked what ailed her. Amba then repeated the entire tale of her misfortune and misery.

Parshuram consoled her, saying, 'I shall send word to Bhishm. I have not the least doubt that he will do whatever I ask him to do. But, in the unlikely event of his disobeying me, I shall challenge him to fight me and most certainly, if that were to happen, I should slay him.'

Amba pleaded, 'I was rejected by Bhishm because I had already pledged myself to Shalv. Shalv also rejected me because he entertained doubts about my chastity. You can judge the rights and wrongs of the case in your supreme wisdom. As you see, Bhishm is the root cause of my misery. He is ambitious, arrogant, mean, and you should teach him a lesson. I beseech you, my lord—slay him.'

Parshuram told Amba that he had taken a vow not to resort to the force of arms except to protect the interests of a Brahmin or with his opponent's express consent. 'Bhishm and Shalv will do whatever I ask of them. So, I shall strive, by every peaceful means, to remove your suffering.'

'My child', Parshuram remonstrated, 'Bhishm is a man of virtue, deserving of respect. If I ask him to put his head to your feet, he will readily do so.'

Amba responded to this by saying, 'If you really wish to help me, you must slay Bhishm. You must fulfil your promise to me.'

Akritvran, at this juncture, interposed, 'Sire, do not abandon this poor unhappy girl who has sought your protection. If you merely challenge Bhishm, he will admit defeat. This will solve the girl's problem, and you won't break your vow about not taking up arms.'

Parshuram welcomed this suggestion and said he would take Amba with him and go to Bhishm to try and find a peaceful settlement. But if he failed, he would engage Bhishm in battle and kill him.

And so Parshuram, accompanied by Amba and his retinue of holy men, went to Hastinapur to meet Bhishm. Bhishm was delighted to greet Parshuram and made him an offering of a cow. Thereupon, Parshuram addressed him: 'Bhishm, why did you carry away Amba who was committed to another when you had taken a vow of celibacy? And why once you had done this, did you send her away? You have made her an outcaste. In consequence, no one is willing to marry her. So, Shalv, to whom she was already committed, turned her away. I ask you to marry her and let her lead the life of a virtuous woman. It is not just that she should be humiliated in this manner.'

Bhishm refused to obey Parshuram. He replied, 'It is true that I have taken a vow of celibacy, and neither fear nor pity, wealth nor desire will make me depart from the straight path of Kshatriya conduct.'

On hearing Bhishm's response, Parshuram was filled with rage and threatened to kill him, his ministers, and all his courtiers. But Bhishm placed his head on Parshuram's feet, saying that he was Parshuram's disciple and that he had learnt all his skill in arms from him, so it would be impious of him to take up arms against his guru. However, his pleading was of no avail, and Bhishm was constrained to take up Parshuram's challenge.

A long and bitter duel between the Brahmin guru and his Kshatriya disciple then took place. It lasted twenty-three days. At the end of that period neither combatant had prevailed over his opponent, though the moral victory rested with Bhishm because just as he was about to hurl his special death-dealing weapon, he was persuaded to hold his hand and to spare the Brahmin's life. Parshuram lauded Bhishm's valour and his adherence to the precepts of dharma, and said that there was no one in the world who could defeat Bhishm in armed combat.

He then called Amba to him and said, 'Princess, I strove to defeat Bhishm, but I could not prevail over him. I am unable to do anything more. You are free to go wherever you wish, but I again advise you to go to Bhishm to be under his protection.'

Amba categorically declined to compromise, so determined was she to avenge the wrong she had suffered at Bhishm's hands. She declared, 'I shall devise a means of killing Bhishm.'

So, Amba retired to the solitude of the forest and started on a long and arduous course of fasting, prayer and other austerities

calculated to give her the spiritual power she required. She practised the most rigid regime, visiting hermitages and holy places.

After many years, Lord Shiva himself one day appeared
before her. He lauded her single-minded devotion to dharma
and her self-denying austerity over the years. 'Ask for a boon,'
he commanded. Promptly Amba asked for the only desire she
had—the strength and the ability to slay Bhishm, who had
reduced her to such an abject state of being without a husband,
without a home, and without any peaceful enjoyment of life.
Shiva granted the boon she had solicited, saying, 'You will
achieve this in your next life when you will be born as the offspring of King Drupad. You will grow up to be a quick-footed,
youthful warrior of great valour. You will have the capacity to
remember all that has happened to you in this life and you will
slay Bhishm in the course of armed combat with him.'

Amba was satisfied, indeed more than satisfied. She could
not wait to fulfil her over-powering desire for revenge, so she
piled up a huge heap of withered branches and dry wood on
the bank of the Yamuna river and set fire to it. When the pyre
was burning with a mighty flame, she climbed up on top of it
and lying down, cried out loudly, 'I have entered this fire to
slay Bhishm.'

King Drupad had a strong animus against Bhishm. He had
no offspring and was constantly praying for the birth of a son
who would grow up to kill Bhishm. In the course of his prayers
he invoked Lord Shiva, and when the god appeared before him
in person, Drupad begged for the boon of a son who would slay
Bhishm. Shiva granted the boon, saying, 'You will have a
daughter who will later change into a man.'

In course of time, Drupad's queen gave birth to a beautiful
baby girl. Drupad and his queen kept the sex of the child a
close secret. They let it be known that the royal offspring was
a boy. The child was given the name Shikhandi with all the
rituals prescribed for the naming ceremony of a male child.
This deception was carefully maintained and when Shikhandi
reached adolescence, King Drupad arranged to marry her to
Hiranyavarma's daughter.

It was not long before the young bride discovered the truth about Shikhandi's identity and spoke to her father of the fraud practised on her. Hiranyavarma threatened to wage war against Drupad to avenge the wrong done to him and to his young daughter. This plunged Drupad's household into panic.

Shikhandi, seeing herself the cause of so much trouble quietly left the palace, and stole off to the abode of a yaksha. Shikhandi sat under a tree and began a fast. It was not long before the yaksha saw her and asked what ailed her. On hearing her tale of woe, the yaksha offered to change sex with her for a time. 'You can go back to your father's palace,' he said, 'and I shall stay here. When your job is done, we shall change back to our original selves.'

This was soon accomplished and Shikhandi went back to his father's palace as a young man. When Hiranyavarma's messengers came, they were given irrefutable proof of Shikhandi's masculinity. But the yaksha who was now turned into a woman was in trouble, for Kuber, the lord of yakshas, happened to come to the forest. Kuber saw the change in one of his people and began to ask questions. When he learnt what had happened, he pronounced a curse on the errant yaksha that he would always remain a woman. But a moment later, he relented and modified the curse, saying that the yaksha would continue to be a woman during Shikhandi's lifetime, and that upon Shikhandi's death he would revert to his original masculine state.

Drupad now took steps to train Shikhandi in the use of arms and the art of warfare. He sent Shikhandi to become a disciple of Dronacharya, a redoubtable instructor in all aspects and branches of the military arts. Soon, Shikhandi came to be recognized as a brave, fearless, and skilful warrior. And, in addition, he possessed Amba's unappeasable fury of the woman scorned.

It was not long before the battle of Kurukshetra began its eighteen-day long, merciless slaughter. Shikhandi entered the fray at an early stage by making a determined assault on Ashwatthama. The duel between the two warriors was brief but bloody. Shikhandi shot arrow after arrow at his opponent. Ashwatthama, in turn, killed the horses of Shikhandi's chariot. At this,

Shikhandi was forced to retreat and the combat remained undecided.

On the evening of the ninth day of the war, Arjun spoke to his charioteer, Krishna, in this manner: 'Bhishm must inevitably die at Shikhandi's hand. This is the decree of fate. Bhishm knows the truth about Shikhandi because he had employed spies dressed as beggars and old men to keep track of Amba and all she did wherever she went. They kept Bhishm informed of Amba's ascetic life in the forest and how she had vowed to destroy him, how she had been granted the boon of slaying him in her next rebirth and how she would assume male form. Bhishm, who would not raise a hand to strike a woman, knows that Shikhandi is in all reality a woman. So, when Shikhandi attacks him, he will turn his face away. I shall place Shikhandi in front of me and remain behind him. Thus Shikhandi alone will fight with Bhishm, and if any Kaurav warrior attacks Shikhandi, I shall promptly eliminate him.'

Accordingly, the next day, this plan was put into action. Shikhandi, protected by Arjun, fired three arrows targeted at Bhishm's chest. But Bhishm merely laughed, in response saying; 'Go on, I shall not retaliate. I shall not hurt you. Fate brought you into the world as a woman and I shall never strike a woman.' Shikhandi shot more arrows at Bhishm and Arjun kept exhorting his ally to continue his assault. 'You must not relent,' Arjun urged. 'If you leave the battlefield without killing him, people will heap calumnies on you and say you did not redeem your vow.'

So, Shikhandi kept up a veritable barrage of arrows aimed at Bhishm. Bhishm, however, did not aim his arrows at Shikhandi, though he was ruthless in slaughtering the other Pandav warriors on the battlefield, fulfilling his vow to slay ten thousand enemy combatants each day of the war.

But Bhishm suddenly felt he had seen enough of the carnage on Kurukshetra. For ten long days, men had killed each other mercilessly. He was appalled by the utter futility of all the death and destruction. He said to himself, 'I have spent so much of my life in killing my fellow beings. To what end? I am tired of life, and this seems to me the appropriate time to die and to put an end to all this waste.'

Yudhisthir, at this very moment, saw his chance to win the war. He exhorted his men to attack Bhishm. It was planned that Shikhandi should push forward, and with the support of the others, he would be able to finish Bhishm. The battle now became even more fierce and bloody than before. The warriors engaged their foes—chariot against chariot, elephant against elephant, cavalryman against cavalryman, foot-soldier against foot-soldier. Shikhandi advanced, sending a tornado of arrows aimed at Bhishm, while Bhishm shot death-dealing shafts at the men of the Pandav army. While Shikhandi continued his unceasing assault, Bhishm looked sideways at the woman-turned-man, as if he would burn up from the fierceness of his wrathful gaze. But he did not aim a single arrow at him. Arjun, once again, exhorted Shikhandi, 'Don't hesitate or pause to think. Act quickly and kill Bhishm. There is no one else in Yudhisthir's army who can stand up to him, O valiant one, this is the truth.'

Shikhandi renewed his attack and moved forward to give cover to Arjun, to whom Bhishm was turning. Shikhandi fired arrow after arrow at Bhishm. Bhishm then left his chariot. His bow had been smashed, he removed his protective armour and stopped firing arrows at Arjun. Shikhandi then saw his chance and fired ten more arrows at Bhishm. Bhishm fell down prone on the ground with his head towards the east and his feet toward the west.

Bhishm suddenly remembered that his mother Satyawati had granted him two boons—that no one would ever defeat him in armed combat and that he would die only when he willed the end to come. So, now, as he lay on a bed of arrows, he kept his lips tightly closed so that no cry of pain should escape him.

The war went on relentlessly. On the sixteenth day Shikhandi fought a bloody battle with Kritvarun of the Kaurav army. He survived till the eighteenth and final day of the war and then met his end in a most gruesome and treacherous manner.

In this war, there were no rights or wrongs, no rules of honour or dishonour. Though the combatants frequently talked of high ethical ideals and the absolute compulsion to observe the precepts of dharma at all times, both in peace and in war, all that seemed to really matter was to achieve victory, regardless of the means. So, on the night following the eighteenth day of the war, when the Pandavs had retired to their camp and were resting, an ignominious and treacherous deed was enacted. Ashwatthama, accompanied by Kripacharya and Kritvarma, went stealthily to the Pandav camp. Ashwatthama sneaked inside the camp and began slaughtering the Pandav warriors as they lay asleep on their beds. He cut Shikhandi in two with a single mighty blow of his sword. The three conspirators, then set fire to the whole camp.

Thus perished Shikhandi, born Amba. But he, or one may say, she, had no regrets. She had wreaked her vengeance and redeemed her honour.

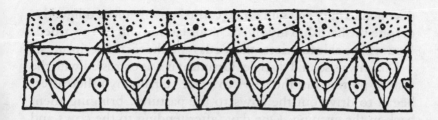

A Guru and His Disciple

A true disciple must always practice the virtues of truthful speech, unswerving devotion, and absolute obedience to his master. Then only can he achieve wisdom, knowledge, and spiritual powers. This precept has been repeatedly asserted and illustrated by numerous stories and instances in the Mahabharat Puran. One such story relates to the sage Dhaumya and his three disciples, Upamanyu, Aruni, and Ved.

One day, Dhaumya asked Aruni to go to the field and see that an adequate quantity of water was flowing in through the irrigation channel. When Aruni reached the field, he saw that a breach in the channel was diverting water away from the crop. He tried to dam the breach, but try as he might, he could not correct the flow. Seeing no easy way of carrying out his guru's directions, he lay down in the channel and thus effectively repaired the breach.

When some days had passed and Aruni remained absent from the hermitage, Dhaumya asked where his disciple had gone. Upamanyu was quick to reply, 'Sir, you yourself sent him to repair the breach in the water channel.'

'Oh!' said the guru, 'Let me go and see what he has been doing all this time.'

On reaching his field, Dhaumya called out Aruni's name and bade him come forward. Hearing his guru calling him, Aruni got up and related how he had dammed the breach in the only possible way. Dhaumya was well-pleased with Aruni's devotion to duty and said, 'You carried out my order diligently, despite the hardship involved. Your labour will be rewarded. You will be able to master the Vedas and the sacred scriptures.'

Dhaumya's second disciple, Upamanyu was entrusted with the task of putting out the cows to graze. Upamanyu thereafter, began to drive out the cows to the pastures, bringing them back in the evening. One day, after tending to the cows and tying them up in the shed, Upamanyu went to salute the guru and stood before him. Noting the robust, indeed, corpulent appearance of the young disciple, Dhaumya asked him what kind of nourishment it was that made him appear so bursting with vitality. Upamanyu replied that he lived on the alms given to him by kind-hearted persons. Dhaumya warned him, 'It is highly improper to consume the alms you receive before placing them before me. This is the homage due to your guru.'

From that day on, Upamanyu began handing over to Dhaumya all the food he received by way of alms. A few days later, Upamanyu again stood before the guru to pay his respects. Dhaumya again noticed that, despite the faithful compliance of his direction, the disciple was no thinner and was, in fact, as robust and corpulent as before. Somewhat surprised, Dhaumya asked him what nourishment he took. Upamanyu replied, 'First, 'I beg for alms for you and then a second time, I beg for myself. I live on the additional food I receive in this manner.'

Dhaumya admonished the young man, 'It is contrary to the sacred precepts to beg for alms a second time. Such greed ill becomes a truly austere and virtuous disciple.' Upamanyu promised never again to beg for additional alms. Upamanyu continued looking after his guru's cows. After some days, he again stood before Dhaumya to make his obeisance. Again, the guru noticed the well fed appearance of the disciple. So, again he asked him what kind of nourishment had contributed to his wellbeing. The young man replied that he was taking sustenance from the milk furnished by the guru's cows.

'But that is not right,' exclaimed the master. 'The cows belong to me. You can't drink their milk without my permission.' Upamanyu submitted to this also. But again when he appeared

before the guru a few days later, the guru again remarked on his far from austere appearance and put the same question to him about the source of his rich nourishment. 'I do not take the cow's milk,' he replied, 'But I imbibe the froth from the mouths of the calves when they suck the teats of their mothers.'

'This is wrong,' said Dhaumya. 'The calves, out of consideration for you, leave much too much froth, and in the process, remain undernourished.' Upamanyu again deferred to his guru's wishes, and promised never again to lick the froth from the mouths of the calves.

Forbidden to ask for alms, drink the milk of the cows, or the froth from the mouths of the calves, Upamanyu looked for other means of sustenance. Finding nothing better than the bitter-tasting leaves of a bush, he began munching them in order to satisfy his hunger. Very soon the poison in the leaves affected his vision. One day, almost blind, he stumbled and fell into a well. When at sundown he had not returned home, Dhaumya said to the other disciple, 'I have repeatedly stopped Upamanyu from what he was doing. It seems that he became angry and has gone away, let us go and look for him.'

On reaching the forest, Dhaumya shouted Upamanyu's name repeatedly and called to him to come.

'I am in the well,' shouted back the disciple and told the guru that he had lost his vision and so had fallen. Dhaumya advised him to seek his salvation by praying to Ashwins and glorifying their virtues and powers.

So, for a long time, Upmanyu chanted the Ashwins' praises and their pious deeds till he was released from the misfortune that had befallen him. Dhaumya, for his part, commended the disciple's devotion to duty and imparted to him the essence of the sacred scriptures. Thus was Upamanyu rewarded for passing the difficult test of the disciple's true veneration of his guru.

The third disciple, Ved, was assigned an equally, if indeed, not more arduous task. He was asked to work as the guru's domestic servant and to tend to his daily, if not hourly needs. Dhaumya made the young man work like a draught bull. He bore the heat of summer, the cold of winter; often hungry or thirsty or in pain, he never complained and looked after every requirement of the sage. He passed the test to the complete satisfaction of Dhaumya and was eventually allowed to take his leave.

Back home, Ved set up his household and adopted three disciples. But he made no demands on their time or leisure. He assigned no duties to them. He could not put away the memories of what he had undergone in his guru's hermitage, and refrained from subjecting his disciples to similar treatment.

But old traditions die hard. When one of Dhaumya's disciples, Uttank, after the conclusion of his pupillage asked him what he should bring by way of a disciple's offering, Dhaumya hesitated, thought over the matter, and finally referred Uttank to his wife. Uttank asked the sage's wife what he should give her as his parting gift. She promptly said she would like the earrings of King Paushya's queen, but that he must hand them to her before the evening of the fourth day, when she had to perform an important sacred ceremony.

Uttank immediately set out for the realm of King Paushya, and after overcoming a number of obstacles because of his determination to accomplish the task undertaken by him, came back to the hermitage with the ear-rings in time for the ritual.

Thus the guru-shishya, master-disciple relationship continued in strict accordance with the prescribed rules and practices laid down by the ancient sages.

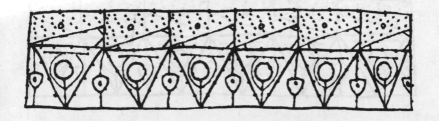

The Wife's Place

King Pratip, the renowned philanthropist, sat on the bank of the river Ganga, performing austerities and meditating on the sacred precepts. The Goddess Ganga, seeing his handsome manliness, was filled with desire to be loved by him. Assuming the form of a most alluring young maiden, she approached Pratip and sat down on his right knee.

Pratip asked her what she wished of him. She replied that she had fallen in love with him and desired him, 'The wise', she argued, 'have always condemned a man who rejects a woman in love.' King Pratip said, 'Beautiful one, I cannot accept another's wife or a woman of different caste. This is my firm resolve.'

Ganga insisted, 'Since my beauty is free of imperfection, I am in no way unworthy of you. There is no caste bar, nor am I another's wife. I am a celestial being. I beseech you to accept me.'

Pratip answered, 'Beautiful one, by sitting on my right knee you have disqualified yourself. This place is meant for a son, a daughter, or a daughter-in-law, not for a wife. So I am unable to concede your wish. It is the left knee which is the rightful place for a spouse or a beloved one.'

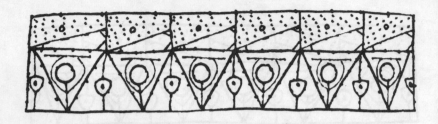

Devyani

The gods and the demons were engaged in a fierce struggle to gain supremacy over the three worlds. The demons were slowly pushing forward and gaining ground, because Shukracharya, their head priest and adviser was able to work a magic charm known as sanjivani, whereby he could bring back to life every demon who was slain by an enemy. Brihaspati, the head priest of the gods, had no matching means of restoring the depleted armies of the gods. So, the gods went to Kacha, the eldest son of their priest and asked him to go to Shukracharya and enter his service as his disciple. 'You are a young man of intelligence and practical sense,' they said. 'You will be able to insinuate yourself into his good graces. Also, Shukracharya has a beautiful daughter, Devyani, on whom he dotes. If your are able to flatter and cajole her and win her heart, you will most certainly be able to wheedle the sanjivani charm from her father, and our purpose will be achieved.'

Kacha accordingly went to Shukracharya and begged the sage to accept him as his disciple. Shukracharya welcomed him and said he would be happy to be his instructor. It was not long before Kacha won the guru's confidence and affection, as well as the young Devyani's. The demons, however, did not relish the presence of someone from the camp of their enemies, and therefore looked upon Kacha with suspicion and hostility. Fearing that Kacha might by some means or other acquire the magic charm sanjivani, they decided to do away with him.

One day, when Kacha was out alone, grazing the acharya's cows, they caught him and killed him. Then they cut up his body into pieces and fed it to the wild beasts. In the evening, the cows returned to their shed, but there was no sign of Kacha. Devyani was alarmed and went to her father. 'The sun has set,' she said to him. 'You have performed your evening ritual. The cows have come back, but there is no sign of Kacha. I feel certain that he has been killed. Believe me, I cannot live without him.'

Shukracharya replied, 'My child, if this be so, I shall at once bring him back to life.' The sage immediately recited the sanjivani charm and Kacha was restored to life. Kacha then came and sat beside Devyani and related to her all that he had undergone.

It was not long before the demons again saw Kacha alone with the cows in the forest grazing grounds. They again killed him, and grinding his body into a fine paste, threw it into the sea. Again, Devyani, not seeing him return with the cows, begged her father to bring him back to life. Once again, the magic of the sanjivani charm recited by Shukracharya brought Kacha back to life, and Devyani regained her beloved companion.

Yet a third time the demons captured and killed Kacha on finding him alone in the forest. This time they burnt his body and mixed the ashes in wine and gave it to Shukracharya, who unwittingly drank it. Kacha's mortal remains thus became part of the sage's own body. This time when Devyani beseeched her father to bring Kacha back alive, Shukracharya advised her not to bestow her affection on a person so prone to die as Kacha. Her charm and beauty were a matter of knowledge and renown in all the three worlds. All the gods and demons, all the worthy Brahmins and indeed, everyone on earth and in heaven was enamoured of her and ready to pay court to her. But Devyani was adamant, and protested that Kacha was no ordinary man, that she had set her heart on him, and that she just could not live without him. Without Kacha, she would abstain from food and drink, thus she would put an end to her sorrow and her life.

Moved by his daughter's grief, Shukracharya called out to Kacha before reciting the sanjivani charm. Kacha answered from inside Shukracharya's own abdomen and related how he had reached this state. Shukracharya was now in a quandary. He was willing to bring Kacha back to life, but in accomplishing this his own body would be torn to pieces.

Kacha suggested a way out of this impasse which would resolve the matter. Shukracharya should first impart the entire secret of the sanjivani charm and the mode of reciting it to Kacha, before actually reciting it to restore Kacha to life. In this way Kacha would have learnt the sanjivani charm before his body became whole, and when Kacha was restored to life and the sage's own body lay torn in pieces, Kacha would recite the charm and make Shukracharya whole and alive again.

So it was agreed, and Kacha achieved his objective without doing harm to his guru or earning his displeasure. Indeed, Shukracharya expressed his satisfaction with Kacha's devotion, as well as his wisdom, and said that Kacha would henceforth be like a son to him. Kacha, for his part, touched Shukracharya's feet and solemnly vowed to treat the guru as he would his father and mother and to always respect him and do his bidding.

Kacha spent some years ministering to his master's needs and wishes, and then prepared to go back to the celestial abode of the gods. Devyani then spoke to him of the deep and unswerving love she bore him. She reminded him that she had not, till then, spoken of her feelings or behaved like a lover because he was under a vow of celibacy while he was her father's disciple, but he must know how she had time and again striven to keep him alive. Now that the period of his rigorous discipline had ended, she was free to declare her love and her deep desire to have him as her husband.

Kacha responded to her, saying, 'Beautiful one, I feel toward you as I would toward a sister. You are most dear to me, but you are my guru's daughter and, in every way, like a sister, as indeed the guru's daughter should in all propriety be.'

Devyani persisted, maintaining that Kacha was not her father's son, and, being the offspring of Brihaspati, he could according to all sacred precepts be her husband. She was irrevocably in love with him, and it was his bounden duty to marry her. Kacha, however, argued that in view of all that had happened, especially his having been reborn out of Shukracharya's own body, he was virtually his son and her brother. He had very deep affection for her and did not wish to displease her, but she must understand his position and the pious motives which guided his conduct.

But Devyani was not appeased, and when Kacha insisted on leaving without her, she pronounced a curse that the knowledge of sanjivani would not give him the power to restore any one to life. Kacha said in reply, 'I declined your offer of marriage not because I saw any fault or flaw in you, but because I regard you as my sister. I submit to the curse you have pronounced. I shall communicate the knowledge of sanjivani to another who will be able to make effective use of it.'

Kacha was welcomed back by the gods, and there was much rejoicing to celebrate the successful accomplishment of his undertaking. It was decided that with the added advantage of the sanjivani charm, the gods should launch a determined assault on the stronghold of the demons and claim a victory. So Indra and his celestial hordes emerged from their realm and with great fanfare started for the neither world. On the way, Indra saw a group of beautiful maidens bathing in the waters of a lake situated in a forest clearing.

He caused a strong wind to blow through the clearing and to scatter the clothes which the young women had placed in a heap before entering the water. When the maidens came out of the water to get dressed, no one could find their own clothes. Each one picked up whatever came to her hand and put it on. In this manner, Sarmishta, the beautiful daughter of Vrish, the head of the demon forces, put on the clothes of Devyani. Devyani was so incensed by this that she accused Sarmishta of committing an unpardonable crime. 'You being the daughter of my father's disciple,' she stormed, 'have dared to wear my clothes.'

Sarmishta answered back with equal acrimony, 'Devyani, are you not ashamed of speaking to me like this? Your father is for ever cringing before my father, lauding him and receiving alms and gifts donated by him as by way of charity to a Brahmin. I am the daughter of one who gives gifts and you are the daughter of one who humbly receives them. I am the daughter of a king, you are only a poor Brahmin, a mendicant. I care nothing for you.'

The quarrel led to a scuffle and Sarmishta dragged Devyani to a dry well nearby and pushed her in. Believing that she had rid herself of her enemy for good, she went home.

But in a short while, King Yayati, the illustrious son of Mahush, who was out on a hunting expedition, happened to come by. He saw the well, and desirous of quenching his thirst, went up to it. When he looked inside, he saw not water but a young maiden, radiant with beauty, standing inside. On being questioned, she told him who she was and how she came to be at the bottom of the well. 'Hold my right hand,' she almost commanded, 'and pull me out.'

Yayati, on being told that she was the daughter of a Brahmin sage, did not hesitate. He took hold of her hand and pulled her out of the well. He then left to continue his hunt. Devyani sat down by the well angry and mortified. When her maidservant came looking for her, she declared that she would not return home till her humiliation was atoned for. 'Go to my father,' she ordered, 'and tell him all that has happened and tell him that I shall not set foot in Vrishaparvan's city till this wrong is righted.'

Shukra, on hearing the maidservant's tale, came running to the forest. He hugged his daughter lovingly and tried to comfort her. 'Everyone', he pleaded, 'receives the reward of his or her deed. You must have done something wrong, which has borne the fruit of suffering.'

Devyani retorted, 'Whatever may be the cause of my suffering, the words uttered by Sarmishta are like thorns in my heart. The demon girl spoke to me in great wrath. With her eyes blazing with rage, she called me the daughter of a mendicant, a recipient of alms, and arrogated to herself the status of one born of a generous demon with royal blood running in his veins and one who is lauded and honoured by all. If it is true that I am

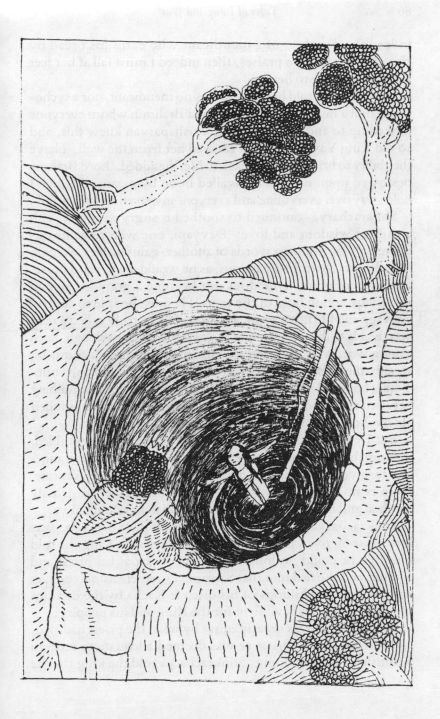

indeed the daughter of a mendicant who earns his bread by chanting her father's praises, then indeed I must fall at her feet and pay homage to her.'

Shukracharya told her that he was no mendicant, nor a sychophant, but a highly gifted and learned Brahmin whom everyone looked up to and revered. King Vrishaparvan knew this, and so did King Yayati who had rescued her from the well. 'I have the ability to bring the dead back to life,' he added. 'Lord Brahma bestowed upon me this unrivalled boon by means of which I hold sway over everything and everyone in heaven and on earth.'

Shukracharya continued to soothe his angry daughter with words of wisdom and love: 'Devyani, one who remains unmoved by the insulting words of another, gains victory over all. He who can control his passions, as he would a wayward horse, attains perfection. One does not become a good horseman merely by taking hold of the reins. He must be able to control a savage and unruly mount. He alone can be truly called a man who is able to discard his anger as a snake discards its slough. Such a one is worthy of acquiring the quadruple gifts of righteousness, wealth, the satisfaction of earthly desires, and spiritual salvation. Young boys and girls are immature and incapable of distinguishing between right and wrong. They call each other names, but their elders should not involve themselves in such disputes.'

But Devyani refused to be appeased. 'Father', she grumbled, 'I know the difference between right and wrong, between anger and forgiveness, but when a disciple does not behave like one, there is no question of forgiveness. I have been so grossly insulted and humiliated that I have no wish to remain alive.'

Shukracharya, seeing his daughter suffer so much, was himself soon overcome by wrath. Going to the King, he told him that he could no longer continue to act as his priest and adviser, and warned him that Sarmishta's reprehensible conduct and the repeated killings of his disciple Kacha by the demons would one day bear evil fruit for the King and his people.

Vrishaparvan was greatly upset by what the priest had said, and offered to do whatever was deemed adequate to redeem Sarmishta's wrong doing. Shukracharya told the king that he could not bear see his daughter suffering, and that the only way of alleviating her pain was for the King to go to her and make amends. The King readily agreed and going to Devyani asked her what would right the wrong done to her.

Devyani replied, 'I wish that Sarmishta, along with a thousand of her hand maidens, should serve me and go with me to whatever places my future husband takes me.'

Vrishaparvan conceded Devyani's demand, and sent a messenger to his daughter conveying his decision. Sarmishta promptly complied with her father's command, and after confessing her fault, declared that because of her insulting behaviour, Shukracharya and Devyani must not leave her father's court. She then took a thousand maidservants and went to Devyani's apartments to place her services and those of her maidservants at Devyani's command.

Devyani, taken by surprise, said; 'How will you, the daughter of a King who is glorified by a poor recipient of alms, act as the slave of the self same alms receiver? Sarmishta replied; 'Be that as it may, I shall go with you as your slave, wherever you go. Thus I can keep my people happy and they can continue to benefit by the advice and presence of the holy and spiritually powerful Brahmin.'

And so, Sarmishta and her thousand maids began to live in Shukracharya's apartments and to minister to the needs of her new mistress, Devyani. Thus days and weeks of happiness passed spent eating delicious viands and drinking wine. One day, Devyani and Sarmishta, accompanied by their maidservants, went to the forest and as they were resting near the dry well into which Sarmishta had pushed Devyani, Yayati again happened to arrive in pursuit of deer. Seeing a crowd of women relaxing in the glade, he stopped and asked them who they were and why they had come to the forest. Devyani in turn asked him who he was and where he was going. Then, remembering her encounter of many months ago, she told him that she had chosen him for her husband and she wished to marry him and serve him faithfully as his legal spouse. She would also have Sarmishta and a thousand maidservants for tending to him and her own needs.

Yayati answered; 'You are Shukracharya's daughter, and I am not worthy to be your husband. Your father will not be willing to give your hand to a Kshatriya. The wise have said that a Brahmin's wrath can destroy whole cities and realms. So, I cannot accept you unless your hand is given to me by your father.'

Devyani replied, 'It is not you who are asking for my hand. Indeed, you held my right hand when you pulled me out of the well, so you have already accepted me as your wife. My father will with certainty, bestow my hand on you formally with the proper rituals.'

Devyani thereupon sent one of the maidservants to fetch her father. He arrived without delay, and when the matter was explained to him, he made a formal offer of his daughter's hand to Yayati. Yayati readily agreed to marry Devyani and begged Shukracharya to absolve him of any wrong resulting from the inter-caste marriage. Shukracharya blessed the union and declared that no blame would be attached to him for marrying a Brahmin girl. He added; 'The demon girl, Sarmishta, will also go with Devyani. Treat her with due consideration but do not make her lie in bed with you.'

Yayati took Devyani and all her retinue to his capital city. He installed Devyani in the palace and provided a suitable residence for Sarmishta in a well-appointed building situated in the centre of a grove of Ashoka tress, not far from his palace. Thus he passed his days in luxury, performing his royal duties and taking his due share of marital bliss. In the course of time, Devyani presented him with a handsome son who lived in the palace apartments and gave endless joy to the parents.

Sarmishta, too, yearned for the love of a husband, a noble and handsome spouse who would bless her with a son. One day when Yayati went to the glade of Ashoka trees, he saw her standing alone, looking so tantalisingly beautiful that he stood speechless. Sarmishta, with joined hands spoke to him. 'Sir, you know that no one can ever go near the living apartments of Indra, Vishnu, Varun or Yama. It is even so in your palace. I am desirous of a suitable spouse so that I may find fulfilment in giving birth to a son. I beseech you to grant me my desire.'

Yayati answered, 'I know how very beautiful you are. Your ancestry and your disposition are above reproach. But Shukra-charya had admonished me not to take you to my bed. So I am helpless.'

Sarmishta replied, 'Listen, Sire, the wise have declared that on some occasions it is permissible, nay prudent, to utter an un-truth, for instance while joking or speaking about a woman who has lain in bed with you, on the occasion of marriage, when death threatens you, or when you may run the risk of losing your fortune. So, it cannot be said that one who utters a false-hood in such an eventuality falls from the status of righteous-ness and virtue.'

Yayati argued that people looked up to the King as a paragon of virtue, and if he told a lie or acted contrary to his sacred pledges and obligations, he would certainly perish, so he was not prepared to prevaricate. Sarmishta went on pleading and arguing. 'A woman's husband', she insisted 'is in every respect equal to the husband of her friend. When you married Devyani, you practically became my husband also.'

Yayati pondered over this statement, looking all the while at the tantalisingly alluring figure standing before him. Finally, he said, 'I have vowed never to turn away one who comes to petition me. Tell me what you want of me.'

Sarmishta repeated her request. 'All I say is save me from straying from the path of virtue to do what is unchaste. If you give me a son, I shall be seeking a righteous aim. Look, a woman, a slave, and a son have no rights over their property, for it really belongs to the husband, the master, or the father. I am Devyani's slave. She rightfully belongs to you. Therefore, she and I are both deserving of your companionship. So do accept me.'

Unable to refute this incontrovertible argument, Yayati agreed to stay with her for a time and fulfil her wish to have a son by him. In the course of time a male child endowed with heavenly beauty was born to Sarmishta.

When Devyani received the news of this event, she was filled with suspicion and jealousy. She immediately sought Sarmishta and asked her who had fathered the child. Sarmishta replied that she had been granted the boon of a son by a learned and pious sage who had visited her.

'If you are telling the truth', said Devyani, 'I am happy, but tell me the name of the sage.'

Sarmishta pleaded ignorance in this respect, and for the time being, saved herself from the jealous wrath of Devyani.

Life went on as normal till some time later Devyani gave birth to another son who matched his brother in looks and accomplishments. Sarmishta, too, received the gift of two more sons. All the five boys grew up happily, playing and romping in their respective homes. One day Devyani and Yayati happened to go to the grove of Ashoka trees near Sarmishta's home, and there they saw three lovely boys playing games. Devyani was taken aback by the sight and wanted to know whose children they were. She looked askance at Yayati, and then asked the boys who their father was. The boys, in reply, pointed their fingers at Yayati and rushed up to embrace him. At the same time, they said that Sarmishta was their mother. Yayati, non plussed and afraid of Devyani, did not reciprocate their affectionate gesture. Instead he tried to free himself from the boys' embrace. Crestfallen and surprised at his sudden coldness, the boys ran away crying and sought the consolation of their mother, who was standing under a tree nearby and watching the scene.

The truth came to Devyani in a flash. Her voice choking with wrath, she lashed out at Sarmishta. 'You are a devil woman. You were my servant. How could you dare deceive me? Aren't you afraid of reverting to your demon ways?'

Calmly and gently, Sarmishta answered, 'I did not lie when I told you that a sage had fathered my son. I have acted justly and in accordance with the rules of virtuous conduct. I have no reason to be afraid of you. When you pledged yourself to be Yayati's wife, I too took a vow that I would have him for my husband. I admit that your status as a Brahmin is higher than mine, but do you not know that Yayati is dearer to me than he can ever be to you.'

Devyani now turned her fury on Yayati and told him that she would never again live with him. She ran weeping to her father and Yayati, with heavy foreboding, followed her. With her eyes burning with anger, she burst out, 'Father, vice has triumphed over virtue. The low have risen and the lofty have fallen. I have been wronged by Sarmishta. She has three sons by Yayati, while I have only two. He was reputed to be a right-eous person, but he has turned out to be an unmitigated villain. He has bestowed a bigger marital gift on Sarmishta than on me.'

Shukracharya did not ask Yayati to explain or defend himself. He immediately passed sentence by pronouncing a curse. 'Since you have wilfully and wickedly transgressed the sacred path of dharma, you will forthwith be afflicted with the scourge of old age and senile decreptitude.'

Yayati tried to justify his conduct by pleading that according to religious precepts, the man who turns away a woman who comes to him with the desire to conceive a son by him is as sinful as he who kills an unborn child in the mother's womb. But the argument did not persuade Shukracharya, and Yayati saw his young and handsome body rapidly become shrivelled and decrepit before his very eyes.

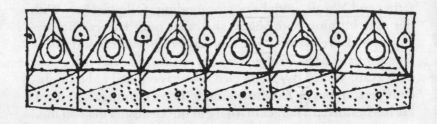

Kaal (or time)

Who can erase the writing of fate and time? Who can undo the
doings of Kaal? What is to be and what is not to be, the joys
and sorrows of life, happen according to the dictates of Kaal.
Kaal causes chaos and Kaal restores order and peace. Kaal nour-
ishes all living things and makes them grow, and Kaal reduces
them to nothing. Kaal creates all things and Kaal destroys
them. Kaal alone is awake when all else is asleep. No one can
overpower Kaal. All events, past, present, and future are
ordained by Kaal.

Remember this and do not let your wisdom be clouded.

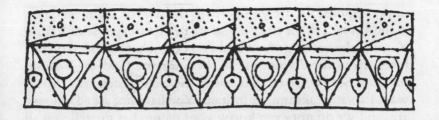

Jayadrath

The Pandavs, during their period of exile, were living in the Kamyak forest. Jayadrath, King of Sindhu, was seeking a bride. His quest led him close to the hermitage in which the five brothers and their newly wedded wife, Draupadi, had taken up their abode. The young ruler was dressed in royal garments and was accompanied by a retinue of minor princes. He struck camp not far from the hermitage. The Pandav brothers had gone on a hunting excursion and Draupadi was standing at the hermitage door and whiling away the time.

Jayadrath saw her from far and was at once bewitched by her breath-taking beauty. He was overpowered by an irresistable desire to possess her and knew that he had to look no further for a bride. He asked his friend, Prince Kotikasya, to go and find out who she was and whether she was already married. 'Also, ask her,' he said, 'if this young woman with a beautiful face, pearly teeth, and slender waist, will accept me.'

Kotikasya descended from his chariot, and going up to Draupadi, addressed her. He was like a jackal approaching a tigress. Kotikasya said, 'O beautiful maiden with arched eye brows holding the branch of a Kadamb tree and glowing like a flame of fire fanned by breezes, how is it that you are standing alone in this hermitage? Your appearance is superior to any other in all the three worlds. Are you not afraid to be alone in this vast and wild forest? Are you the spouse of a god, or a demigod, or are you, by chance, an apsara—a nymph of the

upper regions? Or are you perhaps the wife of a demon living
in the forest? Tell me truly if you are the divine spouse of Varun,
Yamraj, the moon god, or Kuber. Have you descended on this
forest from the divine seat of the Supreme God, of Vishnu,
Indra, or Surya? Or are you one of the celestial denizens, Saraswati, Parvati, Lakshmi, or Indrani? You have not asked who we
are, and we do not even know your name. Let me tell you all
about us, then you in turn can tell me why you have come here.'

Kartik then recounted in detail about himself and about King
Jayadrath whose emissary he was. He dwelt at length on the
glories and wordly possessions of Jayadrath, his vast military
strength and his conquests. Once he had finished he again requested Draupadi to reveal her identity and to say whose daughter
and wife she was.

Draupadi lowered her hand from the branch she had been
holding, and arranging her silk sari, looked at Kartik for a
moment in silence before saying, 'I should not be talking to you
alone in this manner, for to do so is not the approved conduct of
a chaste and virtuous wife. But now I know that you are Kotikasya,
the worthy son of King Surath, and since there is no other man
or woman present to answer your question I am taking the
liberty of speaking to you. I am Krishna, daughter of King
Drupad. You must have heard of the five Pandavs who rule
over the Khandav State, Yudhisthir, Bhim, Arjun, Nakul, and
Sahdev. I am their lawfully wedded wife. My husbands have
gone out to hunt. They must be on their way back, and as soon
as they arrive, they will welcome you and entertain you.'

After saying this, Draupadi went in to prepare something to
serve to the guests.

Kotikasya went and related all this to Jayadrath in the presence of his friends. Jayadrath exclaimed, 'While the beautiful
one was talking to you, I was filled with an insatiable desire for
her. I am surprised that after speaking to her, you came back
empty-handed. There is no equal to her in the whole world.
Verily, from the moment I set eyes upon her ravishing beauty,
all other women seem to me like so many monkeys. Tell me
the truth, is she really human?'

Kotikasya replied, 'She is King Drupad's daughter, Krishna. All the five Pandavs cherish her. Seize her, at once, and take her back to Sindhu.

Jayadrath, now mad with desire and his mind full of evil designs, rushed to see Draupadi personally. As a wolf goes to a tigress, so did Jayadrath enter Draupadi's hermitage. Standing close to her, he began asking her how she was and how the five brave Pandavs fared. Draupadi replied that all was well and she was just preparing something for his breakfast. Jayadrath, now quite impatient, blurted out, 'You have already given me all I need for my breakfast. Get into my chariot and come away with me. You have had enough of this rough and wild life, married to men who have lost their kingdom, their homes, their wealth, happiness, and comfort. Abandon this harsh forest life and be my queen and mistress of the Sindhu state.

Draupadi was filled with rage on hearing Jayadrath's insolent and wicked proposition. Knitting her brows, and moving away, she shouted at him, 'Aren't you ashamed of yourself?, Don't utter another word.'

She then busied herself with household work and began talking of other matters to while away the time before the arrival of one or more of her husbands. But she talked in vain. Jayadrath, now desperate, could brook no further delay. He grabbed her by her clothes and began to drag her towards his chariot. Draupadi resisted with all her might, and pushed him away. Jayadrath fell like a tree which has been cut down, but quickly jumped up and once again grasped the end of her sari. Jayadrath was stronger than Draupadi and very determined. Draupadi, scuffling with her attacker, was dragged and carried to the chariot. At once, the entire procession of princes, courtiers, and soldiers, began moving off out of the forest. All the while, Draupadi heaped curses on her abductor and warned him that his heinous deed would be mercilessly avenged.

It was not long before the Pandavs returned from their hunting expedition. The birds were screeching uneasily and wild animals were moving about restlessly. As they neared the hermitage, they observed the signs that some terrible calamity had occured.

When they came to the hermitage they saw that it had an empty and forlorn appearance. Draupadi's maid servant was weeping and moaning. On seeing her masters, she wiped her eyes and told them that the villainous Jayadrath had forcibly carried away her beloved mistress.

'He can't have gone far', she moaned, 'because the tree branches, broken by the passage of his chariot, are still green and unwithered. Go quickly and rescue the beautiful princess. Put on your armour, take your bows and arrows and go in pursuit of the villain. Rescue her from his hands before he soils her virtuous body . If that happens it will be a terrible tragedy. It will be like a ladle of sacred ghee being poured on a heap of ashes, or throwing a garland of beautiful flowers on the cremation ground. It will be like a dog lapping up the sacred soma juice placed near the holy fire, or a jackal who, after eating discarded flesh, soils the waters of the sacred lake by stepping into it. Let not a low sinner lay his filthy hands on your bright-eyed, beautiful beloved. Go, do not delay another moment.'

Yudhisthir told her to cease her wailing and her harangue. 'Move away,' he commanded, 'don't make such improper and painful suggestions about Draupadi. Anyone who dares do such a senseless thing, be he a king or a prince, will die at our hands and pay for his drunken arrogance.'

The Pandavs followed the forest path indicated by the marks left by Jayadrath's passage, all the while loudly twanging their bowstrings to intimidate their foe. Soon, they saw a cloud of dust in front of them and urging on their chariot steeds, they overtook the foot soldiers of Jayadrath's retinue who were trailing behind his chariot.

When they caught sight of Draupadi on Jayadrath's chariot, they were filled with rage against him. With one voice, they called to him to stop. Their loud stentorian cry rang so fiercely in the forest that Jayadrath lost his nerve and asked Draupadi if the men following them were indeed her five husbands. Draupadi, now sure of being rescued, turned upon her abductor and answered, 'Yes, these brave warriors are the five Pandavs, my husbands. But of what benefit will this knowledge be to you now, since death is about to overtake you for committing such a heinous deed. Fool that you are, you may have the answer to your futile question.'

Soon, the Pandavs overtook Jayadrath and launched an attack on him and his men. Jayadrath fought valiantly and so did his men, but they were no match for the invincible Pandavs who were angered beyond measure by the sight of their captive wife. Bhim, wielding his heavy mace slaughtered whoever came before him. His valiant brothers inflicted fatal blows on all around them. Kotikasya fell a victim to one such blow, so did many other princes. Seeing the disastrous debacle, Jayadrath jumped down from his chariot and ran from the battlefield. His soldiers abandoned the fight and ran helter-skelter into the forest. Bhim was determined to slay the perpetrator of the insult to which Draupadi had been subjected and declared that he would not spare the villain but would follow him to the end of the world to kill him. Yudhisthir counselled moderation, and said that though he was guilty of a heinous crime, his life should be spared because he was related to Gandhari and the Pandav's own sister.

But Draupadi's indignation at what she had suffered could tolerate no half measures. Addressing Bhim and Arjun, she exclaimed, 'If you agree to do what I truly desire, you will not spare the life of this lunatic miscreant. One who lays his hands on a chaste woman or another's kingdom does not deserve mercy even if he begs forgiveness with joined hands.'

At this, Bhim and Arjun set out in search of Jayadrath while Yudhisthir, Nakul, and Sahdev drove Draupadi back to the hermitage. The two warriors urged their steeds in pursuit of Jayadrath, now fleeing for his life. As soon as he came into view, Arjun sent a well-aimed arrow at his horse, instantly killing it. Jayadrath was thrown off, but he quickly got up and ran off. In no time, Bhim overtook him and picking him up bodily, hurled him down with great violence. Then, falling upon his prostrate body, Bhim began to pummel him with his fists and feet. Arjun pleaded, asking Bhim not to kill the wretched man, reminding him of what Yudhisthir had said. Bhim, feeling angry and frustrated, had to be content with shaving off Jayadrath's hair, leaving a few tufts as the stamp of disgrace. Bhim said to him, 'Fool that you are, listen to what I say. If you wish your life to be spared, you must always declare yourself to be the slave of the Pandavs. On this condition only will I spare you.'

Jayadrath had to submit to Bhim's diktat. Bhim and Arjun then secured their vanquished foe with chains, and placing him on their chariot, drove back to the hermitage. They placed him before Yudhisthir and related to him the manner in which they had made him their captive and how he had agreed to be their slave in return for being spared his life.

Yudhisthir saw Jayadrath's pitiable state, his chained, dust-covered body, his shaven head, and the look of utter despair on his face, and could not refrain from laughing. After a moment, he told Jayadrath that he was free to go away and that the condition about his being their slave was waived. 'But', Yudhisthir added on a serious note, 'don't ever again tread such a wicked path, and be careful to act in a truly virtuous manner.'

The once haughty and powerful monarch was thus defeated, humiliated, and disgraced. He could not go back to his country and face his people. Nursing a deep and bitter grudge against the Pandavs, he firmly resolved to take his revenge in equal measure.

Jayadrath took the road to Haridwar. Arriving there, he sat down on the bank of the sacred Ganga and began to pray to Lord Shiva. He started on a rigid course of fasting and prayer and practised the most mortifying austerities in order to fortify his resolve and strengthen his spirit. He continued in this manner till one day Shiva relented and appeared before him in person.

'What do you wish?' asked the Lord from out of whose matted locks the sacred Ganga sprang.

'I pray for a boon,' Jayadrath said.

'What boon?'

'That I may defeat the Pandavs in battle.'

Shiva demurred. 'You cannot overcome the five Pandavs. It is ordained that no one can defeat them or slay them. But I can grant you the power on one occasion to repulse the four of them other than Arjun. As for Arjun, no one, neither god nor man, can prevail over him. Indeed, I myself have invested him with the divine weapon Pashupat.'

Jayadrath had to be content with this inconclusive boon. Thanking Shiva for his qualified help, he went back to his kingdom to wait for the day the boon granted by Shiva would help him take his revenge.

It was a long and distressing wait, but the day finally arrived when the great war at Kurukshetra commenced. The opposing armies were ranged against each other. Jayadrath had joined the Kauravs. It was decided by the Kaurav council of war that he should join battle with Satyaki. Bhishm was of the opinion that Jayadrath, driven by his feelings of unjust treatment at the hands of the five Pandavs and emboldened by Lord Shiva's boon, could be depended upon to acquit himself with honour.

Jayadrath's first encounter with the enemy took place on the very first day of the war, when Drupad, the father-in-law of his foe, sought him out and attacked him. Jayadrath at once shot three arrows at his assailant. Drupad retaliated with a veritable shower of arrows, many of which inflicted deep wounds upon Jayadrath, but the duel remained inconclusive.

It was not till the thirteenth day that the aggrieved ally of the Kauravs was able to display his true valour. The Kauravs' army had taken up the battle formation known as Padmavayuha. Arjun's son Abhimanyu, a young but fearless warrior not yet out of his teens, decided to launch a charge and break up the formation. He pushed through one end and drove his chariot into the very heart of the formation. Jayadrath at once rushed up to stand guard of the breach, thus preventing Abhimanyu's retreat and also making ineffective Yudhisthir's, Bhim's, and Shikhandi's attempt to rescue him. Jayadrath stood firm, resisting the onslaught of the Pandav warriors. This, was in reality the fulfilment of Shiva's boon. Jayadrath was thus able to block the attempt to save Abhimanyu, who died fighting an unequal battle inside the strong army formation.

Arjun now resolved to get even with his old enemy. He took a vow that if he was not able to slay Jayadrath before sunset he

would commit suicide by allowing himself to be consumed by fire. So, on the following day, he continued to seek out Jayadrath and to make him the target of his bow. To prevent Jayadrath from being killed, a number of top-ranking Kaurav warriors made a protective ring around Jayadrath while he sat in his chariot with his head lowered, making himself invisible to anyone outside his circle of protecters. They argued that if in this manner they could keep him alive till sunset, Arjun would be obliged to commit suicide, and if that happened the war could be deemed to virtually have been won.

In the evening, when the sun was moving rapidly to the western horizon, Arjun asked his charioteer, Krishna, to take him to where Jayadrath sat frightened and hunched up in his chariot, his head invisible, waiting hopefully for the sun to disappear from view and anticipating death at any moment. He was surrounded by Duryodhan, Ashwatthama, Kripacharya, Karan, Vaishasen, Shalya and others who formed a formidable and impenetrable barrier. Arjun was getting impatient and his wrath was rising with every passing moment. Krishna told him to wait. 'Look, Jayadrath is protected by six of the bravest warriors. We cannot even see him and you cannot reach him unless you slay all six of his defenders. On such an occasion one must take recourse to trickery. I shall, by my spiritual power, cast a shadow over the sun. When this happens, Jayadrath will think the sun has set. He will be overcome with joy and will fearlessly raise his head, wanting to see you set fire to yourself. That will be the moment. Act quickly and make him the target of a well aimed arrow. Don't think the sun has really set.'

Arjun agreed and so it came to pass. Krishna through his yogic power darkened the face of the sun, Jayadrath, relieved beyond measure and full of joy, raised his head to have a fuller view of the setting sun. 'Look, look,' whispered Krishna to Arjun.'Jayadrath's head is raised. This is the moment.'

Arjun immediately let off a succession of arrows from his quiver, the force of which was so great that Jayadrath's head was severed from his body.

Thus was Arjun's vow redeemed and Draupadi's insult

avenged. The shadow over the sun was then dissolved . For a few moments the setting sun lit up the horrors of the day's carnage before it sank below the horizon.

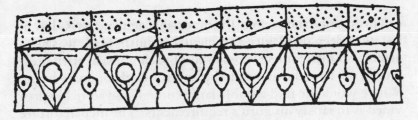

Madhavi

Galav was a devout and conscientious disciple of the great Vishvamitra. He spent many years imbibing wisdom and knowledge from the vast treasure-house of his master, and in return rendered unstinted service to him in the manner of a true and grateful disciple. At length, the time came for him to leave his master's hermitage, though he knew in his innermost heart that there was a great deal more to learn and there was much more that he could do to prove himself a true disciple. The measure of Vishvamitra's wisdom and learning was boundless and Galav's own passion for acquiring spiritual wealth was insatiable, but the burden of dependence had to be relieved so that his master should not be too heavily taxed. So he communicated his resolve to the guru and asked him what he should give by way of his parting homage.

Vishvamitra lauded Galav's dedication to duty, his exemplary aptitude to study, his obedience, devotion, and his selfless service to the needs and wishes of his guru. He added that there was really no need for Galav to offer any worldly gifts to him. Galav, however, insisted on performing a duty which, he was convinced, had not only the sanction of custom but was in the nature of a categorical imperative decreed by the law governing the relationship of guru and shisya. He kept repeating, 'What offering shall I make? Guruji, what offering will be acceptable to you?' till Vishvamitra, somewhat irritated by Galav's obstinacy, said, 'Galav, since you are so stubborn, quickly bring me eight hundred horses, white like the moon, having black ears.'

This strange and seemingly impossible demand put Galav in a state of deep perplexity and mental anguish. How was he to acquire so many horses possessed of such unusual attributes? He did not have the money to buy even a single ordinary horse. Where would he find a stable large enough and so richly provided as to satisfy his guru's requirements? The mental torture which the prospect of failure brought him deprived him of his appetite and sleep. His body began to waste and shrivel till he was reduced to mere skin and bones. His insides seemed to be on fire, his eyes continually filled with tears. 'Woe is me,' he kept repeating, 'from where can I obtain eight hundred white horses with black ears? I have no money and no friends. I have no desire to live. I shall drown myself in the sea or go to some far away lonely spot and end my life. Alas, I am a failure, a pauper without any of the things that make life bearable. And on top of this, I owe a heavy debt to my guru. Ah, one who owes a debt is never happy. One who does not discharge an obligation is quickly destroyed. The guru gave me so much and I vowed to make a disciple's offering to him. This I am unable to do. It is but right that I should hang myself or take a deadly poison and thus seek refuge with the divine Vishnu.'

While he was thus lamenting his misfortune and the sorry state to which he had been reduced in this way, his old friend, Garud, a celestial bird with mighty wings and a fearsome beak and the favourite mount of lord Vishnu, chanced to come down from his heavenly flight. He asked Galav the cause of his sorrow, and when Galav told him his story, he promptly offered to help him. Lord Vishnu, he said, would assuredly bless the undertaking. 'Mount on my back,' said the divine bird, 'and I shall carry you to all the four ends of the earth—east, south, west, and north—or to any other spot where moon-white horses with black ears are likely to be found.'

Garud explained in detail the attributes and virtues of each region. The east is the direction from where the sun god first appears in all his splendour and glory. It is the gateway of the day and of paradise. The lord Indra makes himself manifest in the east. It is here that Agnidev, the fire god, displays his warmth and the intensity of his wrath. To the south dwell thousands of demons. This is the direction in which everyone ultimately travels. Here, wrong-doers receive their just deserts. Even the sun cannot dissipate the darkness of this land, where mighty serpents like Vasuki, Takshak, and Airavati hold sway. The west, to where the sun god retires after each day's labour, is where night and sleep dwell, holding mankind captive for half their lives. It is from here that the rivers and streams arise and rush towards the sea. And finally, in the north there are mines of gold. It is here that the resplendent Shiva and his divine consort, Parvati, reside. It is in the north that the seven rishis and Arundhati are to be found along with countless other wonders, said Garud.

Having listened to Garud's account, Galav chose to go eastward. Mounted on Garud's back, he flew like the wind. He could hardly see or hear anything. He began bemoaning his plight, saying there was no way of securing the eight hundred white horses with black ears, and all he now wished was to end his life. Garud laughed and said, 'You are talking like a man lacking wisdom. Death will not come to you at your behest. Death is all powerful and self-willed. He will seek you when he decides that your time has come. I can take you where the white horses with black ears may be found. I have a friend, King Yayati of the lunar race. He is very virtuous and very wealthy. Let me take you to him. If you beseech him and I support your prayer, maybe he will give you what you want.'

And so, Garud and Galav came down to earth and presented themselves before Yayati. Garud explained the purpose of their visit and related the entire story of Galav's grief and the predicament in which he found himself. He begged Yayati to relieve Galav's distress by giving him a gift of eight hundred white horses with black ears.

Yayati was elated that Galav had come to him begging for alms, and spoke thus to Galav, 'Your arrival here and your request have made me feel happy and truly blessed. O, sinless one, though I am not now possessed of as much wealth as I was formerly, I shall not disappoint you. There is no greater sin than the turning away of a suppliant empty-handed. And as I am committed to follow the path of righteousness, I ask you to take this my daughter, Madhavi. She is capable of procreating four dynasties. The king who can have her for his consort and beget a son by her will be willing to part with not only eight hundred white horses with black ears but even all his royal possessions. The son, born of Madhavi, will be my grandson and I have no greater desire in the world than to be a grandfather.'

Galav and Garud promptly agreed and accepted Yayati's offer. Garud said that Galav's objective had been all but achieved, so he would leave him to complete his task with Madhavi's help. He bid both of them farewell and flew back to his celestial abode.

After some delibration, Galav decided to go to Haryashv, the famous king of Ayodhya. It was known that he had no offspring, and the prospect of there being no son to succeed him was causing him great anxiety.

Once their, Galav explained the purpose of his visit. When Haryashv saw Madhavi he was overwhelmed by her beauty. He observed that she was endowed with all the charms that make a woman a perfect consort. The backs of her hands and feet, her full rounded breasts, her shapely thighs, her uplifted cheeks and eyes, her slender waist and ample hips, her perfect hair, teeth, fingers, and toes were all models of perfection. Her voice, her navel, and her intellect were unequalled in their depth, her palms, the outer corners of her eyes, her tongue, lips, and palate were an alluring red. All these virtues taken together proclaimed her sublime beauty, and it was clear that a son begotten by her would be worthy of ruling over the entire world.

So, Haryashv accepted Galav's proposition and agreed to beget a son by Madhavi, and, in return, give a bounty of white horses with black ears. 'But,' he added, 'I have no more than two hundred such steeds. You are welcome to take any number of ordinary horses of which I have thousands in my stable.'

Haryashv's statement plunged Galav into a state of deep perplexity, but Madhavi promptly relieved his anxiety by telling him that she had been granted the boon of regaining her virginity

after childbirth. 'Take the two hundred horses offered by the king,' she said, 'and let me raise a son to him. You can then take me to another king possessed of the kind of horses you need. In this manner I can give birth to four sons and you can have all the horses you require.'

Haryashv was overjoyed that the matter had been resolved to the satisfaction of every one. Galav took the two hundred horses and left Madhavi with him for an appropriate period. In due course, Madhavi gave birth to a son. Galav came back to claim Madhavi who had reverted to the state of virginity, so that he could go to another king to acquire more white horses with black ears. Madhavi left the luxurious splendour of Haryashv's palace, and accompanied Galav who spoke to her in this manner, 'Be not dejected. I am taking you to Divodas, king of Kashi. He is a very virtuous and powerful ruler. Follow me. We shall meet with equal success in his court.'

It was not long before they reached Kashi, and sought the king's audience. As soon as Galav began relating the story of his promise to his guru, Divodas exclaimed, 'You need say no more. I already know everything. I have been most anxious to take this maiden for raising an heir to my kingdom. You have chosen to come to me in preference to many others and I consider myself most fortunate. But I have no more than two hundred horses of the kind you require. These I am willing to give you, if I am given the gift of a son begotten by this beautiful maiden.'

Galav readily agreed and presented Madhavi to the king as his consort for an appropriate term. The king gave her his love in the manner Surya gave his to Prabhavati, Agni his to Swaha, the Moon his to Rohini, Yama his to Urmila, Varun his to Gauri, Narayan his to Lakshmi, Vashisht his to Arundhati, Satyavan his to Savitri, Chayavan his to Sukanya, and Ram his to Sita. Indeed, he loved her as all celestial and earthly persons have loved their spouses.

In due course, Madhavi gave birth to a handsome son and Galav acquired two hundred more horses. Then, taking Madhavi with him, he resumed his search. So Madhavi left the opulence and comfort of a royal home and restored to the state of virginity once again, followed Galav till they arrived at the city of Bhojpur.

Once there, they presented themselves before king Ushinar. Galav addressed him in this manner, 'Your majesty, this daughter of mine will mother two sons begotten by you and they will match the sun and the moon in their splendour. By way of her dower, you must give me four hundred horses with black ears. I need the horses to complete the offering I owe to my guru. Now you have no son, but in this way you will be blessed with male heirs to your throne. Thus, you will acquire salvation for yourself and your descendants and be saved from the tortures of the nether world which are reserved as the lot of childless men.'

King Ushinar replied that he was able to give only two hundred horses of the kind Galav needed. So he would be happy to raise one son and not two as suggested. He was not one to fritter away his possessions in the pursuit of earthly pleasures and it was his duty to promote the welfare of his subjects.

Ushinar and Madhavi spent a whole month in great happiness conducting themselves in the manner of virtuous people, limiting their pleasures to what their honestly-earned wealth permitted. They visited mountain valleys and waterfalls, gardens and woody groves; they lived in palaces and sat on terraces or relaxed in the pleasant atmosphere of basement parlours. Thus, at the end of the ordained period, Madhavi gave birth to a son whose splendour was comparable to the sun's glory.

Once again, Galav, taking Madhavi with him, set out on his quest. On the way, Garud met him and with a happy laugh and congratulated Galav on the success of his mission. Galav replied that he still lacked two hundred horses and was wondering where he should go next.

Garud, however, advised him not to continue the quest, for now there were no more white horses with black ears available anywhere in the world. 'The sage Richeek asked for King Gadhi's daughter in marriage,' he said. 'Gadhi demanded a thousand black-eared white horses by way of compensation. Richeek agreed. He purchased the horses in the market in Varun's dwelling place, and handed them over to Gadhi. The

marriage of Gadhi's daughter Satyavati and Richeek was duly
solemnized, and Gadhi then performed the sacrificial ritual of
Pundrika, when he gave the thousand horses to the Brahmins
who had attended the ritual. The three kings who have given
you two hundred horses each purchased them from these very
Brahmins. The remaining four hundred were accidentally
drowned in the river Beas when they were being led across the
torrent. So now there is just no way of your being able to
obtain more horses. You should go to Vishvamitra and offer
him the beautiful Madhavi as his spouse, in addition to the six
hundred horses which you have succeeded in obtaining. If you
do this all your worries will be at an end and you will have
accomplished your task in full measure.'

So Galav approached Vishvamitra with this proposition.
The pious guru was overjoyed and exclaimed, 'Galav, why
didn't you bring me this maiden to begin with. I would in that
case, have had four sons to bless me. However, I will gladly
take her as my bride in order to have one son. Leave the six
hundred horses in my hermitage where they can move freely
and graze as they wish.'

Thus, Vishvamitra begot one son by Madhavi whom he
named Ashtak and whom he instructed in all religious and
social matters. He gave Madhavi back to Galav, and Galav, in
turn, took her back to her father, Yayati.

Yayati was delighted beyond measure to have his beloved
daughter back home, and celebrated her return with great pomp
and show. Very soon he began to think of finding a suitable
husband for her, someone whom by virtue of his character,
status, and wordly possessions should be worthy of a celebrated
beauty like Madhavi. But Madhavi rejected the very notion of
marriage. She said she had seen enough of kingly pomp and
luxurious living. She had studied the basic raw nature of all
men, their lust masquerading as love and their longing for a
male offspring which manifested itself as a form of self-loving
transitory affection for a woman who seemed to be of easy
virtue. She categorically refused to take part in any kind of
swayamvar, and left her father's palace to live in the forest

where she subsisted on wild fruits and edible roots. There, she listened to the song of birds and the cries of wild animals. Thus she found peace of mind and release from earthly desires and the false values of kings and queens.

Thus, also, a devout and conscientious disciple was able to discharge his obligation by relying on the arduous and repeated sacrifice of her body by a truly virtuous woman and by exploiting the lust of three illustrious kings and an even more illustrious saint.

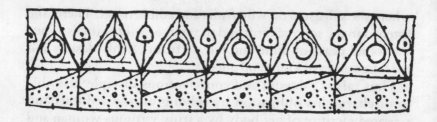

The Good and the Bad
—An Enumeration

Dharma or righteousness is the one supreme virtue, forgiveness the one supreme peace, knowledge the one supreme contentment, and non-violence the one supreme happiness.

A man of high status who pardons and a poor man who gives alms are two beings who live in a region higher than paradise.

Three who are not entitled to own property are the wife, the son, and the slave.

The three gates of hell are lust, anger, and greed.

Four types of men must be shunned by the king: one who is of low understanding, one who is slow in carrying out orders, one who procrastinates, and one who flatters.

These five are deserving of deep respect: father, mother, fire, the soul, and the guru.

These six are prone to forget their benefactors: the student, his guru on the completion of his studies; the son, his mother after his marriage; the husband, his wife after having sex with her; the successful man, his champion after achieving his objective; a passenger, his boatman after crossing the river; and a patient, his doctor after his recovery.

These seven are vices to be abjured by the king: women, gambling, hunting, wine, harsh speech, severe punishment, and acquiring wealth by ignoble means.

These are eight virtues which add to the dignity and lustre of a man: wisdom, high birth, self restraint, knowledge of the sacred books, valour, abstention from loose talk, charitable donations, and gratitude.

The man's body has nine doors: wind, phlegm, and gall are its three pillars; his looks, feelings, the sense of smell, the sense of touch and that of speech are his five witnesses, and over all these presides his soul. He who understands this is indeed wise.

These ten do not comprehend the nature of true virtue: the winebibber, one indifferent to observation, the fanatic, one who is physically exhausted, one blinded by anger, one who is hungry, one who is impatient, one subverted by greed, one overpowered by fear, and one who is lustful.

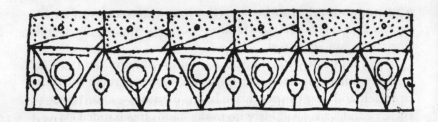

Karan

KUNTI

King Kuntibhoj was a pious, god-fearing ruler. He dispensed benign and even-handed justice to his people, and was magnanimous in extending his hospitality to all who visited his realm. In particular, he was most painstaking in tending to the needs and comfort of any sage or Brahmin who came to beg for alms or to stay as his guest. Being childless, he had adopted Pritha, the daughter of his grandmother's nephew, and had brought her up with love and diligent care. He instructed her in all the sacred precepts and womanly duties. She thus learnt to minister carefully and selflessly to the needs and wishes of the holy men and rishis who came to stay in the royal apartments. She grew up to be a most beautiful, conscientious and dutiful hostess and daughter. She came to be known as Kunti after her father.

One day, a Brahmin of bright and energetic aspect, endowed with powerful limbs and a long shaggy beard and matted locks, came to seek Kuntibhoj's hospitality. He was the learned sage Durvasa. He spoke to the King in a soft voice, full of confidence and dignity and said, 'Peace be to you, Sire. I wish to pass some time here, as your guest, living on your charity. Should you be pleased to accept me as such, I would like to be free to come and go as I wish. No one should hinder my movements. No one may sit on my seat, no one may lie on my bed, or disturb my sleep, during the period of my stay here.'

Kuntibhoj was only too pleased to have the honour of entertaining such a learned and distinguished sage, and readily agreed
to the conditions named by Durvasa. The young princess spared
no pains in discharging her onerous and at times bewildering
duties. She served Durvasa diligently, anticipating his needs
and wishes and catering to all his whims. At the end of his stay,
Durvasa was so well pleased with the care and devotion displayed
by Kunti that in a burst of gratitude he asked her what he should
give her by way of reward for all she had done for him. She
replied shyly, 'Your satisfaction and the affection my father
bestows on me are all I crave for, and I need no other reward.'

Durvasa insisted, 'Though you do not ask me for a gift or a
boon, I, of my own free will, shall teach you a mantra, a secret
charm. By repeating it and naming any god or celestial being,
you will be able to summon the god of your choice. He will
give you a son and carry out your wishes.'

Saying this, Durvasa thanked Kuntibhoj and Kunti for all
they had done for him during his stay as their guest and departed.

Kunti pondered over Durvasa's boon and the charm he had
taught her. She was not a little bewildered by the whole
incident, and wondered if the charm would really bring before
her any god she named. She wanted desperately to put the
charm to test, and yet she was afraid of the consequences.
While she was thus wondering, she felt within her a physical
desire with which for some time she had been periodically
troubled. She pushed away the salacious thoughts induced by
this feeling, but not for long. A day or two later, as she woke
up and lay on her luxurious bed in her portion of the palace
apartments, she saw the sun rising in the eastern sky. She
became so fascinated by the beauty and glory of the colourful
orb resting on the horizon that she continued to gaze at it as it
slowly rose, becoming more resplendent every moment. Yet
her eyes were not dazzled and she thought she saw Suryadev
the sun god himself taking shape, adorned with natural armour
and brilliant ear-rings. The sight fascinated her more and more
till her mind was in a turmoil. She remembered Durvasa's
charm, and on a sudden impulse, she stood up and assuming a

devotional posture—her hands joined together pointing in the direction of the rising sun—she repeated the charm, naming Suryadev. At once, the sun god stood before her in his glowing celestial apparel, his conch-like neck slightly inclined toward her and his honey-coloured eyes looking askance.

'Tell me,' he said, 'what do you wish of me.'

Kunti was so shocked that for a moment she stood stunned, unable to utter a sound. Then she stammered, 'Divine Sir, please go back to where you came from. It was my lack of understanding that made me call you.'

Surya replied, 'If you really mean what you say I shall certainly go back. But it is not right to invoke a god and then send him back. In your heart of hearts you wished to have a son, who would wear armour and ear-rings like mine. Let me beget such a son by you. He will be just the son you long for. I shall then go back to where I came from. But if you refuse, I shall pronounce a curse upon you, upon your father, and upon the sage who taught you the mantra.'

Kunti was in a strange predicament. She was an unmarried virgin, and if she gave birth to a son it would be a terrible calamity—the greatest dishonour that could befall a chaste and virtuous woman. On the other hand, the anger of Suryadev was an even greater calamity. She argued the matter back and forth with her divine visitor, pleading her innocence, her immaturity and lack of wisdom in thus reciting the mantra so early in her life, while Surya insisted on carrying out the purpose of the charm and pointing out the dire consequences of her refusal to accept him. He assured her that after the birth of her son, her virgin status would be restored to her. In the end, she found herself driven to say 'yes' and to submit herself to the sun god's will.

Thereupon, he lightly touched her navel and as she fell back on her bed, he lay down beside her and consummated his purpose. Then, giving her a final parting caress, he went away.

Kunti kept her pregnancy a close secret, known only to her personal maidservant. In due course, she gave birth to a beautiful baby boy, endowed with natural armour and glowing ear-rings. To hide what would certainly be taken as an unpardonable sin and which would be censured by everyone, she placed the child in a box, added some clothes and precious gems, and set if afloat in the stream nearby.

The box was carried down to the river Ganga until it reached Adirath, a charioteer of humble origin belonging to the Soot caste who lived by the side of the river. He saw the box floating on the river and pulled it out. When Adirath saw what the box contained he was filled with astonishment and joy. He carried the child and the contents of the box to his wife Radha. Both of them gloated over their good fortune, as they made plans of how to bring up the child. They gave him the name Vasushen because he had brought them much wealth in the shape of gems and costly ornaments. But he was to become more commonly known as Radheya, after his mother Radha.

PARSHURAM

The child grew up to be a handsome, agile, strong-limbed lad who displayed more of an inclination towards acquiring the military arts in the use of the bow and arrow rather than techniques of horse management or chariot driving. Radheya was for ever practising his archery. He would stand for hours in the sun, praying that he would become a brave and skilful warrior.

By the young age of sixteen, Radheya had acquired uncanny skill in the use of the bow and arrow but he still longed to learn the yet more intricate and advanced knowledge of military strategy from a renowned master. Who, he thought, could be a greater and more competent guru than Parshuram, who had imparted all his knowledge and technique as well as his cornucopia of weaponry to Drona, the preceptor of the Kaurav and Pandav princes? So it was to Parshuram that Radheya went.

Radheya had to be careful to not invite the indignity of being treated as an outcast Soot and, what was even more vital, rejection by his guru on grounds of his low birth. So he donned the apparel and appearance of a Brahmin, and told Parshuram that he was a high caste Brahmin and wished to be his disciple in the study and practice of the military arts. Parshuram accepted the young

aspirant, not suspecting any falsehood or subterfuge, and imparted to him all his knowledge in the use of arms, as well as teaching him the secret mantras by means of which a warrior can become invincible and is able to inflict deadly blows on his foes. He even made him a gift of Brahmastra, the most deadly of all weapons.

Radheya was happy to be so favoured by the greatest of all teachers of the military arts and he spared no pains in rendering every kind of service and devotion to his master. But misfortune was for ever dogging the young man. It so happened that on one occasion when he was very tired, in order to ease his limbs Parshuram put his head on his disciple's lap and went to sleep. Soon, a vicious insect bit Radheya on the thigh, drawing blood. The bite was very painful, but Radheya bore the pain with stoic calm and made not the slightest movement that might disturb the guru's sleep. But when some drops of blood fell on Parshuram's face, he woke up and asked Radheya what had happened. When Radheya confessed that a horrible insect had bitten him but that he had not wished to disturb his guru's sleep, Parshuram exclaimed, 'No Brahmin can calmly suffer so much pain. Of a certainty you are not a Brahmin. Tell me truly who you are.'

Radheya had to confess that he was a low caste Soot and had been brought up in Adirath's house. Parshuram was beside himself with rage at the deception practised upon him, and that in consequence of the deception the knowledge and skill which was the exclusive preserve of the upper castes, the Brahmins and Kshatriyas, had been acquired by a low born Soot. He at once pronounced a curse, that when Radheya's life was in real danger Radheya would forget the charm which had to be recited in order to invoke Brahmastra. Thereafter he dismissed Radheya.

Not long after this mortifying experience, while practising archery in a secluded forest, an arrow shot by Radheya, pierced a calf's body and killed it. The Brahmin, who had owned the calf, was highly angry. He came to Radheya, saying, 'You have killed a Brahmin's calf, thereby committing an unpardonable sin. I

vow that when you will be engaged in mortal combat and
your life is in danger, the wheel of your chariot will fall into a
pot-hole and you will not be able to release it.'

These two curses were to haunt Radheya (or Karan, as he
came to be named after an event that will be presently related).
But they were to make him even more determined to wage
war against his enemies. His inordinate physical strength, his
agility, his extensive knowledge of the martial arts and the
skill in the use of arms that he had acquired as Parshuram's
disciple and, above all, his moral courage and his unbounded
confidence in his capabilities gave him a kind of aggressive
arrogance that brooked no opposition. Radheya felt a deep-set
resentment that fate had been responsible for his being brought
up in the house of a low caste Soot who, though a friend and
supporter of Dhritarashtra, was by his calling only a
charioteer. Radheya felt himself to be far superior to those
who had brought him up. He was endowed with the divine
attributes of ear-rings and a natural body armour which could
protect him against arrows.

THE TOURNAMENT FOR RADHEYA

The first opportunity for Radheya to test his valour presented
itself soon after he had completed his training. Dhritrashtra's
sons had completed their course of instruction in the use of arms
and the study of martial arts. Their guru, Dronacharya, was
well pleased with their performance and he suggested that they
make a public display of their skill. The suggestion was immediately accepted by Dhritrashtra, and a day was fixed for the
exhibition of their mastery in the martial sports. Dronacharya
marked out a suitable arena and had the location adequately
prepared with seating arrangements for the general public as
well as for the élite and the royalty. A separate area was handsomely
furnished to accommodate the queens and princesses and their
women attendants. The tournament was widely publicised by
drums and criers. On the appointed day, a huge crowd of spectators came to watch the Kaurav and Pandav princes (who were all

pupils of Dronacharya) show their skill in the use of arms.

The tournament was formally inaugurated by Dronacharya and the warriors began a display of their arts. After some preliminary bouts, Bhim and Duryodhan walked into the arena and faced each other like two angry male-elephants getting ready to fight for the favours of a she-elephant. They wielded their mighty maces with such masterful skill that the spectators were spell-bound with admiration. However, the combat soon threatened to exceed the bounds of a friendly match and become a truly acrimonious trial of strength. So Dronacharya's son, Ashwatthama, rushed up and separated the contestants to prevent any serious development of the exhibition match.

Drona then announced that Arjun would demonstrate his skill in arms, and went on to say that Arjun, his favourite pupil, was dearer to him than his own son and there was no one equal to him in the use of arms. He could work wonders with his arrows, creating fire, water, and wind in turn. He could hit stationary or moving targets with unerring accuracy.

Arjun then gave a display of his various skills with different kinds of weapons; his swordsmanship, his skills in archery, and his masterful way of wielding the mace.

As he came to the end of his performance and the amphitheatre was resounding to the applause and acclaimation of his admirers, there was heard a discordant noise and a disturbance at the entrance to the arena. Suddenly Radheya, who was by this time known as Karan, made a spirited entrance into the arena wearing his armour and ear-rings and ostentatiously carrying his arms. The valiant son of Suryadev, he was a veritable lion in strength and agility, as powerful as a bull elephant, in appearance bright like the sun, glorious like the moon, and full of energy like a flame of fire. Karan stood looking with disdain at all around him. He declared in a loud voice, 'I am ready to do everything that Arjun has done and much more.'

Everyone wanted to know who the newcomer was. The spectators stood up, agitated by what they saw and heard. Karan, without more ado, began to perform one by one all the acts and exercises that Arjun had displayed. As soon as he had concluded he arrongantly challenged Arjun to a friendly match. But Arjun's temper was up. He wanted to know who this upstart was who had appeared from nowhere and had the temerity to throw a challenge to him, Arjun, the mightiest and the most skilful and accomplished warrior that the world had ever witnessed.

Duryodhan, on the other hand, welcomed Karan and acclaimed the friendship and regard he had always displayed towards him. He declared that Karan was his dear and valued friend. He was ready to share his kingdom with Karan and looked upon him as his superior and overlord. Arjun was beside himself with rage. He felt deeply insulted by Karan's arrogance to which was added Duryodhan's senseless declaration. He turned to Karan and said, 'Those who come to this tournament uninvited and speak in this manner, deserve to die at my hand.'

Karan replied, 'This arena and this tourament are for all, and not just for you alone. Men of royal blood value bravery. They do not, like cowards, utter meaningless threats. Speak with your arrows. I shall presently sever your head with my shafts and let it fall on the ground.'

Thus challenged, Arjun walked up to Karan and stood facing him, ready to begin the deadly duel. There was a sense of foreboding among the spectators, a dread of the unknown outcome of the encounter. The princes in the royal enclosure were silent with fear and anxiety, and Kunti, seeing her two sons prepare to kill one another, swooned. Her maids-in-waiting sprinkled sandalwood water on her face, but she had collapsed in a state of overwhelming fear.

Kripacharya, well versed in the rules governing duels and matches, thought it prudent to intervene, lest what should be a friendly exhibition bout should develop into a disastrous trial of strength between these two heroes, neither of whom could be spared by their respective allies. So, he went up to the spot where the two redoubtable warriors stood facing each other preparing to engage in battle and said to Karan, 'Behold, this is Arjun, a scion of Kaurav blood. He is the son of Pandu, the third son of Kunti. He is ready to joust with you. But you must state your name and the family you belong to. On learning this, Arjun will decide whether to contend with you or not, for princes of royal blood do not measure swords with unknown persons of low caste.'

On hearing these words, Karan bowed his head in shame and confusion. But he remained silent.

However, Duryodhan now responded, 'Kripacharya, men of royal blood, those born into good families, army leaders fit to become kings, if Arjun is unwilling to fight one who lacks a king's status, I, at this moment hereby give Karan dominion over the State of Ang and I invest him with the full status and regalia of a ruling monarch.'

Karan acknowledged Duryodhan's magnanimous gesture and gratefully accepted his new regal standing. In turn, he asked Duryodhan what he could do for him in return. Duryodhan looked at Karan and said, 'I want a deep and lasting friendship with you.' Karan gratefully agreed to Duryodhan's proposition and the two fondly embraced each other to seal the pact.

At this point, Adirath, who had received news of what was happening, made a dramatic entry, supporting his aged limbs with a staff. As he laboured under the stress of his years and anxiety for his son, the cloth covering his shrivelled torso kept slipping off his shoulder, while his entire body was soaked in perspiration. On seeing him, Karan at once rushed up to him and paid him a son's homage by placing his head to Adirath's feet. The old man, embarassed by this public demonstration of filial humility by one who was now a king, quickly covered his feet with the end of his cloth. He embraced his son, joyfully pouring out a flood of tears upon his head.

Bhim now realized that Karan was the son of a humble Soot charioteer. In a voice full of contempt and derision, he addressed Karan, 'Soot-born Radheya, you are not worthy of being slain by Arjun's hand. You are fit only to handle the reins of chariot steeds. You are not worthy of the throne of Ang.'

On hearing these harsh words, Karan's lips began to tremble with wrath. Drawing a deep sigh, he looked up at the sky, searching for the sun.

Duryodhan, filled with rage, his eyes blazing, said to Bhim. 'Bhim Sen, you should not utter such words. A Kshatriya wins respect by the force of his arms. No one knows where either rivers or valiant men are born. Kartikya's origin is hidden in mystery. Some say he was born of fire, others have declared him the son of Kritika, while still others have named him the offspring of Rudra, and some even of Ganga. Vishwamitra—a Kshatriya—achieved the status of a Brahmin. Drona, the unrivalled teacher of the martial arts was born in a pot. Why speak of others? What about the manner in which you Pandav brothers were begotten? This is not hidden from me. Karan, with his natural body armour and divine ear-rings is endowed with every virtue and matchless attributes, he cannot be born of a low-caste woman. Can a doe give birth to a lion? I, his loyal friend, say he is worthy of ruling not only over Ang, but over the whole earth.'

Duryodhan's diatribe fell on deaf ears. Arjun turned his face away. Duryodhan, taking hold of Karan's hand, gently, affectionately, began to lead him out of the arena. Thus the high drama ended. Arjun had given a convincing demonstration of his family pride, an uncompromising awareness of his high birth, and his contempt for the low-born. Karan had manifested an admirable sense of filial duty according to the dictates of dharma. He had, at the same time, suffered a gross humiliation by the taunt hurled at him by Bhim which only deepened the trauma of his previous experience. Duryodhan had cemented a deep and lasting friendship with a brave and brilliant warrior. The day was drawing to its close, the royal attendants were lighting the torches, and the people, stunned by the events they had witnessed, prepared to go home.

THE SWAYAMVAR

As time passed, Duryodhan and his supporters, including Shakuni and Dushasan, brooded over the danger posed by the Pandav brothers to their peace and security. They feared that the five brothers, who were of divine parentage, and who were highly skilled warriors, might try, and fight for the kingdom now being administered by Duryodhan and his brothers under Bhishm's orders. They especially feared Bhim and Arjun. So, they began to devise plans to get rid of their five enemies. Duryodhan, with Karan's approval, sent the Pandav brothers to live in a house constructed of highly inflammable wood and lacquer, with the aim of setting fire to the house and burning them alive, at a later date. But a timely warning by Vidur saved them, and they escaped to a forest where they lived in disguise for some time.

One day, the Pandavs heard that King Drupad was to hold a swayamvar for his beautiful daughter Draupadi, so they decided to compete for the princess's hand.

Drupad had announced that he who could string a mighty bow provided by the king and then shoot four arrows with it, hitting a fish placed above a revolving machine, would win the hand of his daughter. The arrows would have to pass through a hole in the revolving mechanism to pierce the fish.

Many valiant kings and princes, including Duryodhan, Durvishah, Durmukh, Vivinshati, Vikarm, Chitrasen, Vahusali, Shakuni, Sanvala, Ashwatthama, Shivi, and countless others of royal blood, came to try and win Draupadi's hand by the display of their skill and valour. Karan was also among the suitors, as also were Duryodhan's brothers. The five Pandav brothers dressed like Brahmins to conceal their true identity, and went and sat down among the competitors.

Most of the competing princes could not even string the bow, so powerful was its shaft. When many had tried and failed, Karan got up, quickly walked to the bow, picked it up and strung it in one deft movement. He, then, picked up one of the four arrows and placed its end on the taut string. The ease and dexterity with which the whole operation was performed created a sensation, and everyone thought that Karan's arrow would, without a doubt, find the prescribed target. But suddenly, Draupadi shouted 'I shall not marry the son of a low-caste Soot.'

Karan, hurt to the core and mortified beyond endurance, concealed his sense of humiliation and answered Draupadi's insulting words with a superior smile, and flung down the fully strung bow and arrow.

The other kings and princes then tried to perform the task set by Drupad for winning his daughter's hand in marriage. They all failed. Arjun stood up and walked up to the bow. Taking it up, he shot four arrows in quick succession at the prescribed target. All the arrows hit the fish above the revolving machine.

There ensued what was almost a riot: a Brahmin had success-fully performed a feat which all the brave Kshatriyas had dismally failed to achieve. They were ready to fight the Brahmins to prove their valour, and their wrath was most particularly directed at Arjun who had qualified to win the prize of Draupadi's hand. Karan, who had suffered the most, made a direct assault upon Arjun, as he knew that this was Arjun who had once before humiliated and insulted him. But he could not prevail upon his foe, and feeling utterly frustrated and defeated, retired from the unequal contest more determined than ever to get even with the one whom he regarded as his mortal enemy.

KARAN'S VOWS

It was not long before Yudhisthir's passion for gambling was exploited by Duryodhan and his cunning uncle, Shakuni. The Pandavs lost all they had, their kingdom, their wealth, their ornaments and the rich apparel they wore, and were forced to suffer a long period of exile. They set up their living quarters near a secluded lake where they led a life of peaceful poverty, living on whatever the sparse forest provided. They thus began to pass their years of exile. Karan had a strangely vicious trait which was for ever driving him to subjugate and humiliate his sworn enemy, Arjun, and his brothers.

Karan urged Duryodhan to go to the forest and display his might and affluence. 'What happiness,' he argued, 'is greater, than for a prosperous man to see his enemy suffering the pangs of poverty and want? Dress up your queen in rich clothes and let her see Draupadi covering her nakedness with the bark of trees and living in a state of abject poverty.'

Fearing that Dhritrashtra might object to his son vindictively pursuing the Pandavs, for the Pandavs were, after all, Dhritrashtra's brother's sons, Karan pondered over the matter and advised that Duryodhan should go to the forest on the pretext of supervising the management of his cattle and the pasture lands to which they had been taken. Dhritrashtra would certainly raise no objection to such an expedition. And so it was decided, the plan receiving Dhritrashtra's approval. Duryodhan, Shakuni and Karan laughed together and shook each other by the hand at the prospect of seeing the Pandavs thus humiliated.

They went to the forest accompanied by an army of soldiers and a host of women. They took with them 8,000 chariots, 3,000 elephants, 9,000 horses and countless foot-soldiers, carts, a troupe of vaishya girls, traders, hunters, and Brahmins to recite from the scriptures at the appropriate times. They set up their camp a few miles from the humble hut on the lake side where the Pandavs had taken up their residence.

But when they advanced towards the lake, they were met by a multitude of the Gandharvas and their King Chritrasen, who claimed the forest as his domain. Karan, Shakuni, and Duryodhan tried to fight their way through, but the Gandharva army prevailed over them and captured them as their prisoners. Only Karan managed to run from the battlefield to take refuge at a safe spot near Hastinapur.

On hearing what had happened, Bhim was delighted. 'It serves them right', he exclaimed, 'they came here for a wicked purpose, and have been justly dealt with. There is always someone who comes to help the helpless.' But Yudhisthir did not agree with Bhim. He said the captives were their close blood-relations and they must be rescued and freed, by peaceful negotiation, if possible, and failing that, by the use of force.

Bhim and Arjun had ultimately to use force against Chitrasen's men, and the Kauravs were freed. Yudhisthir said to them, 'Brothers, don't ever do this kind of foolish thing again. He who acts rashly from evil intentions is never happy. So go back home with your brothers and cleanse your minds of evil thoughts.'

Duryodhan felt ashamed, and bidding adieu to Yudhisthir,

left for Hastinapur. His pride had been brought low. He felt utterly defeated and humiliated and decided to camp on the way so as to recover from his deep sense of mortification. Karan, who was in a similar state of mind and was hiding not far from Duryodhan's camp, came to him early in the morning and said, "So, you were able to defeat the Gandharvas and leave the battle-field victorious. I was afraid we might lose and be slain or taken prisoner. So I withdrew discretely and have been hiding in a place of safety. You must have seen how the Gandharvas were harassing me and that I was obliged to run away to save myself. It is a matter of great jubilation that you and your brothers were able to defeat the Gandharvas. Truly there is no one to equal you in valour and military skill.'

Duryodhan's eyes filled with tears and he began to disabuse Karan, by telling him how they had been defeated and taken prisoner, and how they had been liberated by Bhim, Arjun, and the Ashwini Kumars, Nakul, and Sahdev.

'Woe is me,' wailed Duryodhan, 'the Chief of the Gandharvas did not hesitate to tell Yudhisthir that we had gone to the forest only to humiliate the Pandavs by looking with contempt at their poverty-striken state. But, be it said to Yudhisthir's credit that when he saw us in captivity and heard of our defeat at the hands of the Gandharvas, he felt sorry and forgave us. There we were, bound hand and foot before our enemies, our royal ladies and the women of the household witness to our disgrace and our abject state as captives. I felt so humiliated, so morti-fied that I now have no desire to return to Hastinapur and to face the accusing looks I shall meet there. I shall stay here and fast unto death. How can I go to King Dhritarashtra? How can I show my face to Bhishm, Drona, Kripacharya and the others? I, who have in happier times placed my foot on the head of my foes and stepped over their bodies am now reduced to this state of degradation. I was led by my folly to do this wicked deed, and I am receiving the due punishment. Oh, I was so puffed up with vanity! I do not deserve to be king. Listen to me, Dushasan, you take my place and be the king. Go and rule over the whole earth, make your friends happy and castigate your enemies.'

Dushasan was shocked by his elder brother's behaviour and his defeatist attitude. He found the composure to say, 'No, no, my brother, this can never be. The earth may split apart, the sky may break into pieces, the sun may cast off its splendour, the moon may shed its coolness, the wind may cease to blow, the fire to burn, the seas may go dry, and the mountains move from their position, but never, never shall I rule any part of the earth without you.'

Karan, who had taken a solemn vow to be Duryodhan's friend and ally to the end of his days, was greatly distressed by Duryodhan's morbid self-pity and urged him to take heart. 'Why are you letting your sorrow get the better of you? Weeping and wailing are of no avail. They do not take away your distress nor do they better your material condition. Strive to be patient. After all, what have the Pandavs achieved? You are still their ruler and they your subjects. It is their bounden duty to serve you and assist you, and that is all they have done. Get up and come home to Indraprastha. Your father and your brothers are waiting to see you. What is the sense of fasting? If you end your life in this ridiculous manner, you will only be a laughing stock.'

But Duryodhan persisted in torturing himself with self-pity, and reiterated his resolve to starve himself to death. He prepared a bed of kusa grass, and clad in old, tattered clothes, sat down to do his penance.

Karan realized that mere words were not enough to rouse Duryodhan's self confidence and pride in his status. The words had to be matched by action. So Karan asked his friend to provide him with men and arms so that he could go and prevail over Duryodhan's enemy. Thus provided, Karan invaded Drupad's territory and was soon able to subdue him. After returning victorious, he arranged a victory sacrifice, and on that occasion took a solemn vow in Duryodhan's presence: 'Until I slay Arjun, I shall not allow anyone to wash my feet. I shall not eat meat, not drink wine. I also declare that until I slay Arjun, I shall not say no to anyone who comes to me seeking alms. I shall hand over to him whatever he asks for.'

The twelve years of the Pandavs' exile were nearly over. It

was expected that they would return to claim their rightful share of the kingdom. It was also widely known that Duryodhan was unwilling to part with a single village to his cousins. Indeed, he looked upon them as enemies to be subdued. His uncle, Shakuni, and Karan, who were his whole-hearted supporters, urged and incited him to fight the five brothers so that they should have unqualified domain over the entire kingdom.

There were frequent meetings and discussions and it was widely spoken of that a war between the cousins was in the offing. Aware of this, Indra wanted to make his son Arjun secure against any danger posed to him by war. Arjun's greatest and most fearful opponent would certainly be Karan, endowed as he was with an impenetrable armour and protective ear-rings. So Indra decided to relieve Karan of these invincible attributes, so that in a man to man contest Arjun's special arms should prove ineffective against Karan. He knew that Karan had taken a vow never to refuse alms to a Brahmin and never say no to any demand made of him, however unreasonable it might be. So, he resolved to go to Karan, disguised as Brahmin and beg for his armour and ear-rings.

That night, as Karan lay asleep in his bed, he dreamt that a Brahmin came to him and admonished him not to part with his armour and ear-rings, no matter who came to ask for them. The Brahmin went on to say, 'These ear-rings and your coat of mail were born with you. Indra, who is always helping the Pandavs will come to you, disguised as a Brahmin, and will ask you to give him both these cherished possessions by way of charity to a holy Brahmin. Do not accede to his prayer. Make any excuse. Offer him anything else but do not give these things which are your body's protective attributes and prevent anyone from inflicting fatal injuries on you.'

The dream Brahmin explained that he was Surya, the sun god and Karan's well-wisher. Karan said to him that he would never say no to a mendicant, and even if his life was asked for by way of alms, he would unhesitatingly yield it. His good name and his reputation as a truthful and virtuous person who never broke his word were of supreme importance to him. The dream Brahmin again advised and admonished Karan not to give away his divine attributes. Once he parted with them, Arjun would be able to inflict deadly wounds on him and would ultimately slay him. 'What good is reputation to a dead man?' Surya asked. 'It is like a garland of flowers on a corpse. Fame is valuable only when you are alive, not when you are dead and your body has been cremated. If you wish to overcome Arjun in armed combat and kill him as you have vowed to do, keep these precious attributes and do not on any account part with them. But if you are determined not to refuse the demand made by the Brahmin mendicant, ask him, ask Indra, for it will be Indra in the guise of a Brahmin, to give you, in return, an unerring invincible weapon—a weapon which when it is aimed at anyone will destroy him. Indra possesses such a Shakti weapon. It has the virtue of destroying anyone at whom it is aimed and then returning to Indra.'

Saying this, the Brahmin disappeared. Karan, when he woke up in the morning, remembered the whole dream with great clarity and pondered over what the Brahmin had said about the virtues of his armour and ear-rings. He decided that should Indra, a Brahmin, or anyone else come and beg for these attributes, he would hand them over, but ask for the Shakti weapon in return.

And so it came to pass. Indra, disguised as a poor Brahmin came to Karan and begged for his armour and ear-rings by way of alms. Karan, at first, offered him unlimited wealth, vast areas of land and anything else he wished for, but Indra persisted and would not deviate from his resolve. Ultimately Karan asked for the Shakti weapon. This Indra promised to give. He said, 'In my hand this weapon is capable of destroying a whole host of enemies, but when you use it, it will destroy just one individual and then come back to me.'

Karan cut off his ear-rings and ripped off his armour and handed it, dripping with blood, to the Brahmin.

BHISHM'S REBUFF

The Pandavs were not without friends and allies. King Drupad and King Virat were beholden to them and were ready to support them in every way. Krishna was also helping and advising them.

Drupad took it upon himself to send an envoy to the Kauravs and to ask for the Pandavs' share of their grandfather's inheritance. When the envoy delivered the message to Dhritarashtra, Bhishm and Karan were also present. While Bhishm welcomed the peaceful demand of what rightly belonged to the Pandavs, Karan became angry and berated Yudhisthir for assuming an arrogant stance because of new strength given to the Pandavs by Drupad and Virat. Bhishm chided Karan at this and reminded him of the time when Karan and six other skilled warriors had been routed by Arjun single-handed.

Ignoring Karan's belligerent attitude, Dhritrashtra sent Sanjay on a peace mission to the Pandavs. Yudhisthir told Sanjay that he was willing to make up with his cousins, provided Duryodhan gave him and his brothers their due share of the kingdom of Indra-prastha. Krishna lent his weight to Yudhisthir's proposal, and added that if Duryodhan refused, a resort to arms was inevitable.

Sanjay went back to Hastinapur and recounted the Pandavs' reply to Duryodhan's plea for a peaceful settlement.

Karan heard what Yudhisthir and Krishna had said and gave vent to an angry outburst, ostensibly to boost the Kaurav morale and reassert his loyalty to Duryodhan. 'I acquired the knowledge of military science from Parshuram by falsely representing myself to be a Brahmin,' said Karan. 'When he discovered my deceit he pronounced a curse upon me, saying that when my end was approaching I would forget the use of Brahmastra and other weapons. I made amends by serving him faithfully and diligently. Listen, Duryodhan, my end is not yet near and I remember all that Parshuram taught me. And I am confident that I can slay Arjun. I will take the responsibility to kill him. I shall most certainly defeat the armies of Panchal, Karush, and Matsya. I shall slay the Pandavs and their offspring, and hand over to you the kingdom I acquire by my victory. Bhishm, Drona, and all the kings who have come as your allies can remain sitting here. I shall go with my army alone to defeat the Pandavs'.

Hearing this boastful harangue, Bhishm scolded Karan,

saying, 'Karan, you have without a doubt taken leave of your senses for you are talking utter nonsense. Though you are talking big, don't you realise that upon your death, all of Dhritarashtra's sons will die. You will curb your arrogance if you only recall to mind the extraordinary valour of Bhim and Krishna when the Khandav forest was burning. You will never be able to use the divine weapon which Indra bestowed upon you. Krishna's deadly, circular, sharp-edged Chakra will smash it to pieces. The weapon you keep worshipping and garlanding will be destroyed by Arjun's arrows. Krishna is Arjun's ally and protector and he will slay you and all his other enemies.'

Bhishm's rebuff touched Karan to the raw. He retorted angrily, 'What Bhishm says about Krishna is true, but to his harsh words about me I have only one answer: I shall not take up arms and fight in this war as long as Bhishm is alive. Only, when Bhishm has fallen on the battlefield and is out of this bloody contest will my deeds of bravery be witnessed by all those assembled here.' With an arrogent flourish, Karan got up, and leaving the assembly, went home.

NEGOTIATION FOR PEACE

The efforts towards avoiding war and in seeking a peaceful solution were not abandoned, despite Duryodhan's categorical declaration, backed up by Shakuni, Dushasan, and Karan, that not an inch of territory would be conceded to the Pandavs, and the Pandav stand that unless they were given their due share of the ancestral estate, they would be compelled to take recourse to arms. Dhritarashtra and Krishna met and discussed the matter. Krishna spoke to Vidur who was as always fair-minded and peace-loving. Krishna then met Karan and told him that he, Karan, was Kunti's kanin or virgin-born son and therefore a maternal brother of the Pandavs. He should not, therefore, regard the Pandavs as his enemies and should make peace with them. But Karan flung back the retort that Kunti had abandoned him as soon as he was born. 'She did nothing,' Karan continued with bitterness, 'to nourish or protect me. I was picked up by

Adirath and brought up by him and his wife Radha. I look upon then as my father and mother. Duryodhan befriended me, and for thirteen years I have enjoyed his affection and generosity. I have ruled over the kingdom he bestowed upon me. How, then, can I abandon him now and go over to his enemies?'

Karan was firm that he would remain Duryodhan's ally and fight the Pandavs, even though they were his own brothers. Krishna returned home a sad and disappointed man. Equally sad and frustrated was Vidur, who had seen the senseless obstinacy of Duryodhan and Karan. He spoke to Kunti about his fears and apprehensions. Kunti resolved to go and speak to Karan and to try and dissuade him from his tragic course of action.

Kunti sought Karan out and found him standing and praying to his father Surya the sun god. When his prayers had ended, he turned round to face her.

'What do you wish of me?' he asked simply.

'You were not born into a Soot family.' She answered. 'You are my kanin, my virgin-born son. Yudhisthir, Bhim, and Arjun are your brothers. Go and join them. Yudhisthir will treat you like his elder brother, and hand over to you the kingdom he will receive from the Kaurav usurpers. I wish and pray that you and Arjun would love each other like Krishna and Balram. There is nothing that you two brave brothers could not achieve together. I beg you to join your five brothers and to live in peace and luxury. I do not want people to go on calling you Soot-born.'

Karan replied, 'It should be my sacred duty to obey you, my mother, but I cannot. You forsook me as soon as I came into the world. You not only took away my good name, you even deprived me of my very life. I was born a Kshatriya, but because of you, I could not enjoy any of the Kshatriya privileges. Even an enemy could not have done me more harm. Now, forgetting the past, you come to me out of purely selfish motives and ask me to give up Duryodhan's alliance. You did not treat me like a son and now for your own good you are talking of the mother–son relationship. And who is not afraid of Arjun and Krishna? If, at this juncture, I join the Pandavs by saying they are my brothers, everyone will say that I am doing it out of cowardice. No one knows of my relationship with them and if now, when war is about to begin, I go over to the Pandavs by declaring my brotherly relationship, what will the Kshatriya warriors say? Duryodhan and his brothers have bestowed upon me their affection and the means of good living for all these years. How can I abandon them now? The time has come for those who were enjoying Duryodhan's protection and material benefits to show their gratitude and to do their duty.'

Karan paused. Looking full into his mother's face, he continued, 'Let me tell you truthfully that I shall fight against your sons with all my might. I deem this to be my bounden, sacred duty. So, even though your admonition is laudable, I cannot act upon it. But your coming to see me will not be in vain. I solemnly promise that in this coming war I shall not slay Yudhisthir, Bhim, Nakul, or Sahdev. By slaying Arjun alone my purpose will be achieved. If on the other hand, Arjun succeeds in killing me, I shall win glory and a place in paradise. And so you will still have five sons, whether I kill Arjun or Arjun kills me.'

Kunti accepted this as her son's final and irrevocable answer. Trembling with fear and anxiety, she embraced Karan and pressing him to her bosom mumbled, 'My son, what you say is just and proper. Don't forget your promise to spare your four brothers, all except Arjun. May you remain free from blame. And now I must leave you.'

THE BATTLE OF KURUKSHETRA

Events now moved fast. All attempts to devise a peaceful settlement had failed, and both sides prepared for war. Bhishm was chosen by the Kauravs as their Commander-in-Chief. Karan had taken umbrage at Bhishm's taunts and had vehemently declared that he would not take an active part in the war as long as Bhishm was alive. Bhishm had openly said that he did not see eye to eye with Karan who, with his hasty temper and endless bragging, was hardly the person to be given any responsibility in the conduct of hostilities. So he quietly accepted Karan's self-abnegating pronouncement.

From the very beginning, the war proved to be a bitter and bloody conflict. Each day many valiant soldiers met tragic ends. Bhishm claimed that he destroyed ten thousand men of the Pandav forces each day. The Pandavs concentrated their attack on the old patriarch, and on the tenth day, he fell, pierced by so many arrows that no part of his body remained whole and unaffected. Indeed, he lay on the battlefield resting on the ends of arrows which had entered his body in countless numbers. He was not dead for he had been granted a divine boon that his soul would leave his body only when he wished to die. But he was out of the combat, and as far as the war was concerned it was as if he were dead. So, Karan could enter the fray without breaking his vow, and from the eleventh day onward he played a true Kshatriya role in the conflict. He prevailed over many a brave Pandav hero and became a veritable terror for his foes on the battlefield. One of his greatest achievements was his triumph over the demon son of Bhim, Ghatotkach, who had caused havoc in the Kaurav forces by his death-dealing tactics, his superhuman physical strength, and the variety of the weapons at his command. Karan was forced by necessity to eliminate Ghatotkach even though in doing so he had to resort to the use of the Shakti weapon, given to him by Indra in lieu of Karan's ear-rings and armour. The Shakti weapon slew Ghatotkach and flew back to Indra in his celestial abode.

This happened on the fourteenth day of the war. That evening, the Kaurav elders called a council of war to plan their future strategy. They unanimously elected Karan the Commander-in-Chief of the united forces fighting for Duryodhan. Karan accepted this responsibility and the honour that went with it He was ceremoniously installed in his new office with due ob servance of the rites and formalities prescribed by the Vedas

Karan now strove to harass and demoralize the Pandavs all the means in his power. But by the evening of the sixteer day the Pandavs seemed to have an edge over the forces un his command. The Kauravs, fearing another night-long stru to the death, decided to declare a temporary truce and returned to their camp.

In the camp, Karan reviewed the situation and declared his intention to make an all out assault on Arjun, who bore the main responsibility for the Kaurav set-back during the day. The next morning Karan spoke to Duryodhan, comparing his own martial strength and skill with Arjun's.

'Look, Duryodhan', he said, 'I shall today fight with Arjun. Either I shall slay him or perish in the attempt. I shall not leave the battlefield till I have killed Arjun or I have been killed. Many of our leading warriors have lost their lives, and I am without the Shakti weapon, but I am the Commander-in-Chief and Arjun must respond to my challenge. Let me share a secret with you. Arjun's weapons and mine have divine attributes. We are both highly skilled warriors, but in the matter of forestalling an enemy's move, in my swiftness of action, in aiming an arrow at a far off target, and in the dexterity of deploying the various weapons, Arjun is not my equal. Also, in terms of my strength and in the effective use of the war machine, I am superior to Arjun. My bow is better that Arjun's Gandiv. With this bow of mine, Parshuram overpowered the Kshatriyas no less than twenty-one times. Listen, Duryodhan, today I shall most certainly kill Arjun and make you and your brothers happy. By the day's end, this realm will be without rivals. It will be your exclusive empire. But I must draw your attention to a most important matter. In one respect, I am less than Arjun and at a disadvantage. He has Krishna to drive his chariot; Krishna to whom everyone pays homage. Also Arjun's chariot is invested with gold embellishments which were gifted by Agnidev, the fire god, and these cannot be destroyed by the ordinary weapons of war, and lastly, the horses harnessed to his chariot are very sure of foot. Nevertheless, I am ready to engage him in single combat. But I wish to have the brave Shalya as my charioteer. If he assumes charge of my chariot, I can guarantee our victory. Shalya, like Krishna, is a very skilful horseman. So this is my earnest desire. If Shalya acts as my charioteer you will witness my valiant deeds, and I shall destroy the entire Pandav army single-handed.'

SHALYA

Duryodhan went to Shalya and conveyed Karan's request to him and begged him, humbly and respectfully, to accede to Karan's proposal. He supported his request by a long and detailed account of Karan's capabilities and his achievements. He pressed upon him the necessity of furnishing him the utmost help in his struggle against the Pandavs and, in particular, in his endeavour to defeat Arjun.

Shalya listened to Duryodhan with rising temper, and when the long homage and the catalogue of Karan's virtues came to an end, he got up. His brow knit with supressed wrath, his eyes blood red and his body trembling, he exploded, 'Listen Duryodhan, your asking me to act as Karan's charioteer in this thoughtless manner is a gross insult. You think Karan is more valiant than I am, and you are lauding him to the skies. Let me tell you, I do not deem him even my equal. I am capable of slaying an entire army single-handed. Just look at the powerful muscles of my arms, my terrifying bow, my murderous arrows, the fast steeds yoked to my chariot, the awe-inspiring chariot itself, and this, my mighty mace. I can, if I so wish, tear the earth apart by my own strength, I can bring down mountains, I can drain off the waters of the ocean, and yet you have the temerity, the insolence to ask me to act as the charioteer of a low-born Soot. It is highly improper of you to suggest such a thing. I shall never agree to obey the orders of a low caste individual on the battlefield. It is written in the Vedas that Brahma's mouth gave birth to the Brahmins, his arms produced the Kshatriyas, the Vaishas were born of his thighs, and the Shudras of his feet. The Shudras, or the Soots, were ordained to serve the Kshatriyas and minister to their needs. No one has ever seen or heard of a Kshatriya serving under a Soot. I was born into a royal family. I was installed on my throne and crowned King. People honour me and render service and homage to me. As such, I can never agree to be the Soot-born Karan's charioteer. Duryodhan, do not insult me, let me go home.'

Shalya stood up and began walking out of the assembly of

kings and leaders of the army, but Duryodhan stopped him, and with folded hands, begged him in all humility to stay. He spoke to him in soft, conciliatory tones. 'What you have said is true. There is not the slightest doubt about it. But in asking you to drive Karan's chariot, I did not mean to insult you. Let me tell you in all humility and honesty that Karan is in no way superior to you in courage or fighting ability. Nor do I have any reservations about your valour. Remember that you promised to assist me in every way. People consider you superior to Krishna in the science of managing horses, and in recognizing their breeding habits, and their capabilities. It is for this reason, and from a desire to win this war with your help and co-operation, that I beg you to take charge of Karan's chariot.'

Shalya, visibly moved by this appeal, replied, 'I am, indeed, happy that you have in the presence of all these royal warriors pronounced me superior to Krishna. I shall certainly drive Karan's chariot. But I want Karan to promise that he will not take objection to whatever I might say when I am driving his chariot.'

Duryodhan and Karan readily accepted this condition.

The morning of the seventeenth day of the great war began with the forces on both sides realigning themselves and making an all out endeavour to gain victory. Karan was in absolute command of the Kaurav forces. His wish to have Shalya for his charioteer had been granted and he was full of confidence in his valour and his ability to devise the correct strategy for achieving the success he had hoped and striven for. However, deep down in his heart there lay some lurking apprehensions. For he knew this to be his day of reckoning. He was determined to slay his sworn enemy, Arjun, or to perish in the attempt. He was not a little troubled by the memory of the two curses pronounced by two Brahmins—one whom he had deceived by lying and the other whom he had offended by mistakenly killing his calf. But he remained undaunted and uninhibited in his utterances, proclaiming his might and his superhuman skill in the use of arms. He sought comfort in the thought that his end was not yet near, for he clearly remembered the charm taught him by Parshuram—the charm which, on account of his guru's curse, would fail to come to his mind when his end was approaching. So, he mounted his chariot and took his seat behind Shalya with the supreme confidence born of arrogance and his determination to wash away the taunts, slights, and the humiliating calumnies hurled at him and his origins and upbringing.

As soon as his chariot approached the battlefield, Karan began to declaim in a loud voice, 'Listen! Tell me where Arjun is. I shall reward the person who shows me where he is with whatever he asks for. If that is not enough, I shall give him a cart loaded with precious gems. If that doesn't satisfy him, I shall give him a hundred cows and for each cow a bronze milking utensil. I shall make him a present of a golden chariot and six richly caparisoned elephants. I shall even give him a whole group of beautiful sixteen-year old maidens, adorned with gold ornaments, who are able to sing and play on musical instruments. If that is not enough, I shall reward him with a thousand swift steeds and four hundred milch cows, each with her calf. I shall add six hundred well trained elephants, each bearing a golden howdah, and decorated with gold and pearl necklaces, together with mahouts skilled in training elephants. I shall give away all my wealth and possessions save only my wife and son. I shall give all this to anyone who shows me Arjun and Krishna. I shall then kill both of them and give away their wealth also.'

Shalya laughed derisively, 'O Soot born, don't plan to give a reward of six elephants with gold ornaments or an equal number of chariots and their oxen. You will see Arjun presently. In your ignorance you are rashly ready to distribute wealth as if you were Kuber himself. You need not make such an effort for you will see Arjun very soon. Do you not understand the wrong done by one who gives away wealth without reason? It would be far better for you to perform sacrifices or to do other good deeds with your wealth. In your blind anger you wish to slay Krishna and Arjun, but I have till today not heard of a jackal slaying a lion. Karan, you are planning to do something which cannot be done. It seems to me that you have no well-wisher. You are jumping into the fire and if you had a well-wisher, he would have stopped you. I think your end is near for you have no understanding of what should be done and what should not be done. One who wishes to live will never utter the kind of follies you have been repeating. The thing you are planning to do is like trying to swim across the sea with a stone tied round your neck or to jump off the peak of a mountain. If you wish to remain alive you should take the help of your army to fight Arjun. Karan, I do not say this out of malice but for your own good and for the good of Duryodhan. Do as I say if you wish to live.'

Karan replied, 'I am looking for Arjun to fight him with my own arms. You pretend to be my friend but speak like a foe, and are trying to frighten me. At this juncture even Indra armed with his thunderbolt in hand cannot dissuade me from my resolve.'

Shalya irritated Karan even more by saying, 'When Arjun's arrows start destroying the Kaurav army and begin coming at you, you will regret having spoken in this manner. Just as an infant lying in his mother's lap wishes to take hold of the moon, you, sitting comfortably in your chariot wish to seize hold of the valiant Arjun. You are a fool. To try to fight Arjun is like running Shiva's trishul over your limbs. It is to summon Yama, the god of death. You think it is child's play to duel with Arjun! Just as a young doe tries to fight with an angry lion you are ready to challenge Arjun. You are like a jackal who, after eating his meal, starts challenging a lion. Or you are behaving like a rabbit who challenges an elephant, or a fool who incites a serpent by prodding him with a stick. You wish to cross an angry sea full of murderous animals. You are making feeble noises like a frog trying to frighten a thundering cloud. Like a jackal, living among rabbits, who thinks he is a lion till he encounters a real lion, you will think you are a lion until you meet Arjun on the battlefield. You will talk proudly till you hear the fearsome twang of Arjun's bow, Gandiv. When you hear Arjun blowing on his blood-curdling conch and see Arjun pull the string of his bow, you will run away like a jackal with its tail between its legs. There is as much difference between you and Arjun as between a mouse and a cat, or a dog and a tiger, or a rabbit and an elephant.'

Karan was beside himself with rage, as he heard Shalya's long and insolent diatribe. He retorted, 'Shalya, only a virtuous man can recognize another possessed of virtue. One who is not endowed with merit cannot see the good in another. And you have always been devoid of virtue, so how can you see any virtue in another. Shalya, I know and appreciate Arjun's divine weapons, his valour, his mighty bow and deadly arrows more than you ever can. I know Arjun's brave deeds and he knows mine. Knowing all this, I am calling out to fight Arjun. I have this beautiful blood-thirsty weapon. I have for a long time kept it protected in sandalwood powder and have paid it due homage. This weapon is capable of destroying a multitude of men, horses and elephants. It can penetrate an armour of mail and smash the bones of its victim. With this I can destroy even the Meru mountain. I am speaking the truth when I say that I shall not release this weapon against anyone save Arjun and Krishna. You will see both of them dying at my hand. In the battlefield I shall slay both of them and then slay you and your brother also. You profess to be my friend and supporter, and yet you are trying to frighten and threaten me with the names of Arjun and Krishna. I know my power and am not afraid of them, so keep quiet. Single-handed I can slay a thousand Krishnas and a hundred Arjuns.'

'Let me tell you what the wise and learned people have said about your country, Madh,' Karan continued. 'And, after listening to me, either keep quiet or make an adequate reply. They say that the people of Madh are deceitful. They are jealous of others. They are low, evil-minded liars and dishonest in their dealings. They have all kinds of evil habits. They are promiscuous in their relationships with women and guilty of committing incest. They eat and drink immoderately. They laugh pointlessly, weep without cause, they sing filthy songs, and indulge in sex indiscriminately. I would have killed you for trying to dissuade me from fighting Arjun, but I have spared you for three reasons. I have to perform my friend Duryodhan's task, in the second place I have promised to forgive you for anything you say and in the third place, I shall be severely criticized if I kill you, an ally of Duryodhan. It is because of these three considerations that I have let you remain alive. But if you utter such nonsense again, I shall smash your head with this mace which is as mighty as a thunderbolt. The kings and princes on the battlefield will witness the death of Krishna and Arjun or see me perish. And now say nothing more but drive my chariot straight to where Arjun is.'

Shalya replied, 'Soot-born, I am not the one to turn away from the battlefield. I was born into a family of kings, and I know my duty. You seem to me like a drunken man. I wish to befriend you and bring you back to your sense. Don't behave like the crow who thought himself a swan's equal.'

Karan said, 'Look, I am not afraid of Arjun, but I am afraid of Parshuram's curse that I shall forget the manner of using the Brahma weapon when my life is in real danger, and also of the Brahmin's curse that, at the critical moment, my chariot wheels will become stuck in deep mud. Your warnings are so much nonsense. Don't you know that you come from a part of the world which is the worst, the filthiest, the most wicked and most sinful of all. So be quiet, or I shall first slay you along with your army and your sons, and then deal with Arjun. Just do your job and move to where the Pandav brothers have stationed themselves.'

THE BATTLE BETWEEN KARAN AND ARJUN

Karan advanced towards Yudhisthir, slaughtering the Pandav soldiers he encountered on his way. He engaged Yudhisthir and quickly defeated him, but remembering the promise he had made to Kunti, spared his life. Yudhisthir, subdued and ashamed, retired from the fray. Karan then attacked Bhim. A fierce battle took place and both combatants sustained injuries. A well-aimed arrow shot by Bhim hit Karan in his chest and he fainted. Shalya, seeing him lying unconscious in the chariot, quickly drove away to a safe distance. But Karan recovered and came back to continue the bloody duel. Karan cut down Bhim's flag from its pole on the chariot and killed the charioteer.

Bhim jumped down and began slaughtering the Kaurav soldiers. Karan, once again, sought out Yudhisthir and killed his charioteer. Once again, Yudhisthir ran away from the encounter, with Karan chasing him. Bhim then ran to the rescue, and Karan turned round to face Bhim and pour all his wrath with vengeful vigour upon him. Seeing this sudden turn of events, Krishna and Arjun made a hurried consultation to devise a plan to slay their most dangerous foe, Arjun repeating his vow to kill Karan.

Karan now attacked his five brothers in turn, and succeeded in destroying their chariots. The Pandavs fought back valiantly, and though they could not vanquish Karan, they slew countless Kaurav soldiers.

At last, Karan moved up to where Arjun was battling against the Kaurav men. Seeing the manner in which Arjun was parrying the blows aimed at him and eliminating his assailants one by one, Karan said to Shalya that Arjun was indeed the bravest fighter of all and truly skilled in the use of arms, adding that he was well assisted by Krishna. 'But,' he said, 'I am determined to engage him and kill him or be killed in the process'.

At this juncture, Ashwatthama made yet another appeal to Duryodhan to make peace with the Pandavs and to put an end to the bloody conflict that had taken the toll of so many brave warriors on both sides and which seemed destined to destroy many more. But Duryodhan was adamant and replied that he was depending upon Karan to kill all the five Pandav brothers to win the war. Duryodhan then went on to urge his men to make a determined advance against their foe and to slay the Pandavs and their allies.

The final round of the combat between Karan and Arjun now commenced. Karan received many injuries. His entire body was covered with blood from arrow wounds. Karan strove to keep himself erect and shot several deadly arrows at Arjun, but there is no escape from the decree of fate. Death was at last ready to lay his hands on Karan. Suddenly, Karan saw that the left-hand wheel of his chariot had fallen into a depression in the ground so that the chariot could not be moved.

At the same time, Karan realized that he just could not recall the magic spell, the mantra, taught by Parshuram. In desperation, he jumped down from the chariot and putting both his hands to the jammed wheel tried to lift it, but he could not.

Seeing Arjun watching him as he strove to free the locked wheel, Karan asked his adversary to hold back his arms till his chariot could be moved. He pleaded, 'Wait, till I have freed my chariot wheel. To attack me while I am engaged in this manner would be the act of a coward or a low-caste fellow. Arjun, you are known to be an honourable warrior. Those who observe the Kshatriya code of war do not attack a man who is praying, or who comes seeking asylum, or is unarmed, or is leaving the battlefield, or whose hair is undone, or who is a Brahmin, or who stands before his opponents with folded hands, or who has no more arrows left, or whose armour is split, or whose weapon had fallen from his hand. Pandav, you are the bravest warrior on earth. You are upright and virtuous; you know the rules of combat and are skilled in the use of all kinds of weapons; you are reputed to have a saintly disposition. Just give me enough time to release my chariot wheel, for you are on your chariot and I am standing upon the ground. It is not proper for you to attack me in this condition. I say this not out of fear. Indeed, I am not at all afraid of you or of Krishna. I am making this appeal to you in the name of righteousness and fair play.'

It was not Arjun, but Krishna who answered Karan. 'O son of Radha, it is indeed fine for you to talk of righteousness at this moment. We know how low-born men start blaming fate when they are in trouble. They forget their own misdeeds. The Pandavs have always acted according to the dictates of dharma. That is why they are now being helped by dharma. Do you remember, Karan, how Draupadi, wearing but a single garment during her menstrual period, was dragged into the assembly of men with your full approval and at your instigation? Were you acting according to the precepts of dharma? When the villainous Shakuni, upon your advice won everything from Yudhisthir by unfair means, where was your dharma then?

When, with your approval and advice, Duryodhan administered poison to Bhim, had snakes bite him, and made every effort to kill him, where then was your dharma? When the house of lacquer in which the Pandav brothers were sleeping was set on fire, where then was your dharma? When you insulted Draupadi and heaped abuses upon her and watched while the Kauravs insulted and ill- treated her, when out of greed, you called the Pandavs a second time to make Yudhisthir lose all he had in a gambling game, where then was your dharma? When contrary to all rules of fair play, you and a large group of your allies surrounded young Abhimanyu and slaughtered him, where then was your dharma? On all these occasions, you did not observe the precepts of dharma, so why are you now shouting dharma, dharma, dharma? You can cry dharma a hundred thousand times, it will be of no avail. This is your time of reckoning. Just as Nala recovered his kingdom after losing it to his brother in gambling, the Pandavs will now recover their kingdom from the wicked and deceitful Kauravs.'

Krishna's invective made Karan feel ashamed. He lowered his head and was silent. His lips were trembling with rage, but then he picked up his bow and began to fight Arjun.

Krishna immediately urged Arjun to shoot the deadliest of his arrows and to put an end to Karan. Krishna's recount of all his humiliations at Karan's hands had enraged Arjun. Arjun fitted the chosen arrow to the string of Gandiv, his unerring bow. He pulled the string till his thumb and finger holding the arrow end reached the lobe of his ear, and taking careful aim, released the deadly shaft. He saw the arrow enter Karan's throat and sever his head, which fell bleeding to the ground. Karan's body lay punctured and bleeding from countless arrow wounds.

It was late afternoon on the seventeenth day of the fratricidal war. Karan's end left no doubt about the ultimate victory of the Pandavs.

ARMY MAKE UP

1 chariot, 1 elephant, 3 horses and 5 footsoliders comprise a *Patti*.

3 pattis comprise a *Senamukh*.

3 senamukhs comprise a *Gulm*.

3 gulms comprise a *Gan*.

3 gans comprise a *Vahini*.

3 vahinis comprise a *Pritna*.

3 pritnas comprise a *Chamu*.

3 chamus comprise an *Anikini*.

10 anikinis comprise an *Akshauhini*.

So one *Akshauhini* equals:

21,870	chariots,
21,870	elephants,
1,09,350	foot soldiers,
65,610	horses, or equestrian soldiers.

THE SEVEN SACRED RIVERS

The seven sacred rivers of ancient Bharat, were: the Ganga, Yamuna, Saraswati, Vitashtha, Sarayu, Gomti and Gandaki.

The Saraswati dried up and its bed is today not visible. It was said to join the Ganga and Yamuna at their confluence near Allahabad at the spot known as 'Triveni'.

Vitashtha is the old name of the Jhelum.

TWELVE TYPES OF SON
(In descending order of quality)

1. Auras: born of wife, begotten by husband, lawfully married to her.
2. Praneet: begotten by a distinguished person of one's wife.
3. Parikreet: born of a married woman, by someone other than the husband, who has been paid money for the service rendered.
4. Paunarbhav: born of a widow by one not her husband.
5. Kaneen: born of a married woman before her marriage.
6. Kund: born of an unchaste wife by a person of her choice.
7. Dattak: a son by adoption.
8. Kreet: son bought with money.
9. Upkreet: son brought up by foster parent.
10. Voluntary: one who comes voluntarily and says, 'I am your son,' and is accepted.
11. Gyatiretasahod: begotten by brother or relative of the husband.
12. Heenyonidhrit: born of a woman of low caste.